AF751506

Carbon Sink Estimation and Evaluation of Ecological Function Changes on the Tibetan Plateau

Wei Zhou · Qi Chen · Qiquan Li · Wenping Yu ·
Yuechen Li · Weiyu Shi

Carbon Sink Estimation and Evaluation of Ecological Function Changes on the Tibetan Plateau

Springer

Wei Zhou
Chongqing Jinfo Mountain Karst
Ecosystem National Observation
and Research Station, School
of Geographical Sciences
Southwest University
Chongqing, China

Qiquan Li
College of Resources
Sichuan Agricultural University
Chengdu, China

Yuechen Li
Chongqing Jinfo Mountain Karst
Ecosystem National Observation
and Research Station, School
of Geographical Sciences
Southwest University
Chongqing, China

Qi Chen
School of Resources and Environment
University of Electronic Science
and Technology of China
Chengdu, China

Wenping Yu
Yibin Academy of Southwest University
Yibin, Sichuan, China

Weiyu Shi
Chongqing Jinfo Mountain Karst
Ecosystem National Observation
and Research Station, School
of Geographical Sciences
Southwest University
Chongqing, China

ISBN 978-981-95-8663-9 ISBN 978-981-95-8664-6 (eBook)
https://doi.org/10.1007/978-981-95-8664-6

© The Editor(s) (if applicable) and The Author(s), under exclusive license to Springer Nature Singapore Pte Ltd. 2026

This work is subject to copyright. All rights are solely and exclusively licensed by the Publisher, whether the whole or part of the material is concerned, specifically the rights of translation, reprinting, reuse of illustrations, recitation, broadcasting, reproduction on microfilms or in any other physical way, and transmission or information storage and retrieval, electronic adaptation, computer software, or by similar or dissimilar methodology now known or hereafter developed.
The use of general descriptive names, registered names, trademarks, service marks, etc. in this publication does not imply, even in the absence of a specific statement, that such names are exempt from the relevant protective laws and regulations and therefore free for general use.
The publisher, the authors and the editors are safe to assume that the advice and information in this book are believed to be true and accurate at the date of publication. Neither the publisher nor the authors or the editors give a warranty, expressed or implied, with respect to the material contained herein or for any errors or omissions that may have been made. The publisher remains neutral with regard to jurisdictional claims in published maps and institutional affiliations.

This Springer imprint is published by the registered company Springer Nature Singapore Pte Ltd.
The registered company address is: 152 Beach Road, #21-01/04 Gateway East, Singapore 189721, Singapore

If disposing of this product, please recycle the paper.

Acknowledgments This work was supported by the National Natural Science Foundation of China (No. 42471355), the Fundamental Research Funds for the Central Universities (No. SWU-KT24012, No. SWU-KF25023), the National Natural Science Foundation of China (No. U2244216), the Sichuan Science and Technology Program (No. 2023NSFSC1916), and the New Chongqing Youth Innovation Talent Program (No. CSTB2025YITP-QCRCX0092).

Competing Interests The authors have no competing interests to declare that are relevant to the content of this manuscript.

Authorship Contribution Statement

Wei Zhou involved in conceptualization, supervision, writing of the original draft, writing of the review and editing, funding acquisition, and methodology. **Qi Chen** participated in data curation, methodology, software, analysis, and writing of the original draft. **Qiquan Li** involved in data curation, investigation, and supervision. **Wenping Yu** contributed to methodology, validation, and funding acquisition. **Yuechen Li** involved in investigation and funding acquisition. **Weiyu Shi** contributed to investigation and supervision.

Contents

Abbreviations

AGB	Aboveground Biomass
AK	Available Potassium
ALOS	Advanced Land Observing Satellite
AN	Available Nitrogen
ANN	Artificial Neural Network
AP	Available Phosphorus
AUC	Area Under Curve
AVHRR	Advanced Very High Resolution Radiometer
BD	Bulk Density
BGB	Belowground Biomass
CASA	Carnegie-Ames-Stanford Approach
CR	Continuum Removal
CV	Coefficient of Variation
DEM	Digital Elevation Model
DNDC	DeNitrification-DeComposition
DNN	Deep Learning Neural Networks
ERA5	ECMWF Reanalysis v5
ESV	Ecosystem Service Value
EVI	Enhanced Vegetation Index
FD	First-order Differentiation
FPAR	Fraction of Photosynthetically Active Radiation
GBDT	Gradient Boosting Decision Tree
GEE	Google Earth Engine
GIMMS	Global Inventory Modeling and Mapping Studies
GPP	Gross Primary Productivity
GWR	Geographically Weighted Regression
HASM	High Accuracy Surface Modeling
HHL	Harrow-Hassidim-Lloyd Algorithm
HJ-1A	Huanjing-1A Satellite
HSI	Hyperspectral Imager
HWSD	Harmonized World Soil Database

IGBP	International Geosphere-Biosphere Programme
InVEST	Integrated Valuation of Ecosystem Services and Trade-Offs
LAI	Leaf Area Index
LST	Land Surface Temperature
LUCC	Land Use and Cover Change
LUE (ε)	Light Use Efficiency
MAE	Mean Absolute Error
MD	Mahalanobis Distance
MESMA	Multiple Endmember Spectral Mixture Analysis
ML	Machine Learning
MLR	Multiple Linear Regression
MODIS	Moderate-resolution Imaging Spectroradiometer
MSC	Multiple Scattering Correction
MSE	Mean Squared Error
NASA	National Aeronautics and Space Administration
NDVI	Normalized Difference Vegetation Index
NEP	Net Ecosystem Productivity
NIR	Near-Infrared
NOAA	National Oceanic and Atmospheric Administration
NPP	Net Primary Productivity
OA	Overall Accuracy
OOB	Out-of-Bag Error
PA	Producer's Accuracy
PAWC	Plant Available Water Content
PCA	Principal Component Analysis
PLS / PLSR	Partial Least Squares (Regression)
R^2	Coefficient of Determination
RF	Random Forest
RMSE	Root Mean Square Error
RPD	Relative Prediction Deviation
RVI	Ratio Vegetation Index
SAM	Spectral Angle Mapping
SAR	Synthetic Aperture Radar
SAVI	Soil-Adjusted Vegetation Index
SD	Standard Deviation/Second-order Differentiation
SDRmax	Maximum Sediment Delivery Ratio
SG	Savitzky-Golay
SID	Spectral Information Divergence
SMA	Spectral Mixture Analysis
SOC	Soil Organic Carbon
SOM	Soil Organic Matter
SVM/SVR	Support Vector Machine (Regression)
SWIR	Short-Wave Infrared
TK	Total Potassium
TN	Total Nitrogen

TP	Total Phosphorus
TRSR	Three-Rivers Source Region
TWI	Topographic Wetness Index
UA	User's Accuracy
VIF	Variance Inflation Factor
XGBoost	Extreme Gradient Boosting
ZY1E	ZY-1 02E Satellite

Chapter 1
Overview

Abstract Soil organic carbon (SOC) constitutes the largest carbon reservoir within terrestrial ecosystems, even minor fluctuations in SOC can precipitate abrupt changes in CO_2 concentrations, exerting profound impacts on global climate change and the carbon cycle. Vegetation biomass serves as a crucial parameter for assessing vegetation carbon sinks and is a key indicator for evaluating regional or global carbon sequestration potential. Consequently, precise estimation of carbon in vegetation biomass is fundamental to quantifying terrestrial carbon sequestration potential. Meanwhile, the dynamics of vegetation and soil carbon sinks constitute the carbon source-sink functions of ecosystems. Furthermore, ecological changes such as the terrestrial carbon sink and vegetation community of Tibetan Plateau and its response to climate change and human activities represent a focal point in global change research. Thus, how have the vegetation biomass, soil organic carbon, vegetation net primary productivity, grassland degradation, and ecological service functions of the Tibetan Plateau changed in the context of increasing global climate change and human activity interference? The above research and driving mechanism explore has important guiding significance for the precise implementation of ecological restoration strategies and the increase of land carbon sequestration for the Tibetan Plateau.

Keywords Soil organic carbon · Vegetation biomass · Data fusion · Machine learning

1.1 Factors Influencing of SOC Spatial Variation

Soil organic carbon (SOC) constitutes the largest carbon reservoir within terrestrial ecosystems, being two to three times greater than atmospheric or vegetation carbon pools (Lal, 2004).Even minor fluctuations in SOC can precipitate abrupt changes in CO_2 concentrations, exerting profound impacts on global climate change and the carbon cycle (Arora et al., 2013; Bradford et al., 2016). Its dynamic variations also constrain crop growth and food security (Zhao et al., 2018). Consequently, there is

© The Author(s), under exclusive license to Springer Nature Singapore Pte Ltd. 2026
W. Zhou et al., *Carbon Sink Estimation and Evaluation of Ecological Function Changes on the Tibetan Plateau*, https://doi.org/10.1007/978-981-95-8664-6_1

an urgent need to accurately capture the spatiotemporal dynamics of SOC to provide essential support for future actions toward enhancing terrestrial carbon sinks and achieving carbon neutrality. Vegetation plays a pivotal role in terrestrial ecosystems, significantly influencing terrestrial carbon cycling and carbon budget balance. With a carbon pool reaching 500 GtC, it constitutes a substantial component of terrestrial carbon storage (Lal, 2004).

Spatial variability is an inherent characteristic of soil (Filippi et al., 2018). As a crucial soil property, the spatial variation of soil organic carbon (SOC) is influenced by complex and diverse factors, with dominant drivers exhibiting scale dependency (Biswas et al., 2013). At large regional scales, factors such as parent material, climate, geomorphological units, and soil types determine spatial variation in SOC (Calzolari et al., 2021; Chaopricha & Marín-Spiotta, 2014; Wiaux et al., 2014). Conversely, at finer scales, SOC distribution are more influenced by topography, vegetation cover, and land use (Li et al., 2020; Mishra et al., 2010). Some research found that the explanatory power of surface soil organic carbon spatial variation will be increased with the scale become more finer (Xie et al., 2004). For China's natural vegetation, temperate grasslands, and coniferous forests, annual mean temperature, precipitation, and altitude explained 38%, 52%, and 53.8% of the spatial variation in surface soil organic carbon, respectively (Xie et al., 2004). And other research found at the provincial scale that soil texture and annual mean temperature explained 32% and 23.4% of the spatial variation in SOC in Jiangsu Province, respectively. Meanwhile research identified that temperature, precipitation, and clay content collectively contributed 49.9% to the spatial variation of surface SOC in dryland farming areas of Northeast China (Sun et al., 2019). However, elevation exerts a dominant influence on the spatial distribution of SOC in mountainous regions (Zhang et al., 2021a, 2021b), while differing aspect within the same forest type also leads to variations in SOC content and spatial distribution (Beatriz et al., 2016). Therefore, the dominant factors for spatial variation of SOC differ with the change of research scale.

From the perspective of soil genesis factors, the five major factors (parent material, topography, climate, biology, and time) are the dominant determinants of SOC levels and spatial variation. The equilibrium between carbon inputs from litter decomposition and rhizosphere deposition, and carbon outputs during decomposition processes, determines soil carbon storage (Zhou et al., 2002). In the context of SOC variability analysis, conventional methodologies that depend on soil sampling and field surveys have historically served as the primary means of data acquisition. However, in recent years, with the advancement of remote sensing technology, satellite remote sensing data has found increasingly widespread application in SOC research. The relationship between environmental covariates and soil organic carbon content is contingent on environmental conditions, spatial resolution, and the extent of the relevant area (Lamichhane et al., 2019). The integration of remotely sensed data on climatic factors, topography, soil types, vegetation, and human activities such as land-use change and population density, geostatistical methods and machine-learning techniques, facilitates the estimation of regional-scale spatial distribution of SOC. When combined with statistical modeling, the contribution rate of driving factors to spatial variability for soil organic carbon.

The influence of climatic factors on SOC content demonstrates significant variation under differing climatic conditions. In humid climates, high precipitation and elevated temperatures promote net primary productivity in vegetation, thereby increasing the quantity of litter entering the soil and thus organic matter input (Sutfin et al., 2016), which promotes SOC accumulation. Furthermore, the presence of moisture in the environment has been shown to enhance the stability of organic matter, thereby slowing the rate of decomposition and increasing the content of SOC (Liu et al., 2019). Conversely, for regions with severe soil erosion, low and unstable rainfall, low biomass productivity and poor soil water retention capacity result in comparatively lower SOC levels (Fu et al., 2010). In arid climates, vegetation growth is constrained, thereby reducing organic matter inputs. Concurrently, insufficient soil moisture accelerates organic decomposition, resulting in lower SOC content. However, at smaller spatial, climatic factors may be disregarded, as the spatial distribution of SOC content in micro-regions is primarily influenced by variations in topography and landform conditions.

Topographic factors constitute the most critical element among the five major soil-forming factors and represent the most commonly employed environmental variables for predicting SOC (Behrens et al., 2014). The impact of elevation, slope gradient, aspect, and topographic complexity on soil organic carbon content is predominantly influenced by indirect mechanisms, including the effects on soil temperature and moisture, the vertical distribution of vegetation, soil temperature and moisture conditions, and soil erosion processes (Zhang et al., 2021a, 2021b). It has been demonstrated that higher elevations generally exhibit lower temperatures, higher humidity, slower organic matter decomposition rates, and consequently higher SOC content (Sinoga et al., 2012), whereas lower elevations show the opposite pattern. As slope steepness increases, there is an attendant increase in soil loss through erosion (Li et al., 2016). Furthermore, variations in slope aspect have been demonstrated to influence soil temperature and moisture conditions, thereby affecting the distribution of soil organic carbon (Bangroo et al., 2020). The aforementioned topographic factors collectively contribute to the pronounced spatial heterogeneity of SOC within mountain ecosystems.

The influence of vegetation types on the input of biomass carbon into the soil and the vertical distribution of SOC content has been demonstrated (Jobbágy & Jackson, 2000). The influence of soil properties on soil organic carbon content is well-documented, with the physical and chemical characteristics of the soil directly governing the microbial processes involved in decomposition (Zhang et al., 2021a, 2021b). The primary source of soil organic carbon is the decomposition of plant residues, roots, and microorganisms. Consequently, soil texture (comprising sand, silt, loam, clay, and mud, etc.) and pH value directly influence SOC content and distribution. The distinct characteristics exhibited by different soil textures have been demonstrated to affect soil organic carbon content and distribution. Sandy soils characteristically contain larger particles, which leads to increased porosity and drainage capacity. Such soils characteristically exhibit a lower organic carbon content, attributable to their unfavorable pore structure, which hinders the accumulation and stabilization of organic matter. In contrast, clay and loam soils contain a high number of fine particles and micropores, which facilitate easier accumulation and stabilization of organic matter within these minute voids (Wagner et al., 2007). Consequently, these soil types characteristically exhibit higher organic carbon concentrations. Furthermore, the rate of decomposition of soil organic carbon is influenced by soil pH. In the context of acidic soils, the rate of SOC decomposition is retarded, with microbial activity being inhibited, yet its content tends to be elevated (Cheng et al., 2013).

Among human-induced disturbances, land-use conversion represents a significant driver of changes in soil organic carbon stocks. Initiatives such as the reversion of agricultural land to forests and grasslands (Hong et al., 2020; Liu et al., 2018) have been shown to enhance soil organic carbon accumulation by increasing vegetation cover. This approach first elevates carbon inputs to the topsoil, improves soil moisture and nutrient conditions, boosts soil fertility, and thereby promotes SOC accumulation. Consequently, the abandonment of sloping cultivated land has been shown to favor the accumulation of surface soil organic carbon (Lan, 2021). Furthermore, scientifically sound agricultural management practices, such as reduced tillage, the retention of crop residue, and the judicious application of fertilizer, have been demonstrated to contribute to increased SOC content (Karunaratne et al., 2014). Conversely, practices such as frequent tillage have been demonstrated to disrupt soil aggregates, resulting in the loss of soil organic carbon (Chen et al., 2017). Soil aggregates are defined as structural units formed by the binding of soil particles and organic matter. They play a crucial role in the stable storage of soil organic carbon. Frequent tillage has been demonstrated to compromise the stability of soil aggregates, resulting in the loosening of soil structure and the acceleration of organic carbon oxidation and decomposition, thereby contributing to its loss. Concurrently, frequent tillage disrupts soil bio-communities, thereby reducing the abundance and diversity of soil microorganisms and soil fauna. This, in turn, impacts the formation and transformation processes of soil organic carbon.

1.2 Data-Driven SOC Estimation

Multi-source remote sensing has enriched the characterization of soil formation environment covariates and is widely applied in acquiring large-scale spatial information on soil organic carbon both domestically and internationally. Hyperspectral remote sensing data, in particular, has seen extensive use in SOC inversion over bare soil areas. Hyperspectral remote sensing provides crucial spectral information for soil property inversion; for instance, utilizing airborne hyperspectral remote sensing (Gholizadeh et al., 2018), satellite hyperspectral data such as Hyperion, EnMAP (Castaldi et al., 2019), and Gaofen-5 (Meng et al., 2020) can be employed to retrieve SOC in small-scale bare soil agricultural fields. In the context of vegetated areas, sensors are unable to directly detect soil information. However, the interactive mechanisms between vegetation and soil provide a theoretical foundation for the use of optical remote sensing data in the inference of soil properties in vegetated areas (Demattê et al., 2017). These interactive mechanisms include the influence of vegetation on soil biogeochemical processes and the converse, i.e., the constraining effect of soil properties on vegetation distribution (Ballabio et al., 2012). Examples of such indices include remote sensing vegetation indices and characteristic bands (Garnier et al., 2022), which have been successfully applied in the production of global SoilGrid products (Hengl et al., 2017) and the Chinese Soil Information Grid datasets (Liu et al., 2022). SAR data in microwave remote sensing, particularly Sentinel-1 polarimetric data, have been demonstrated to effectively capture the relationship between vegetation and soil. This has resulted in significant attention being given to SOC inversion and distribution mapping (Poggio & Gimona, 2017; Zhou et al., 2020). It is acknowledged that alternative biophysical indices, which have been demonstrated to be effective in characterizing vegetation conditions (Mandal et al., 2020), may also be utilized in the context of remote sensing inversion of soil organic carbon content. A plethora of open-source remote sensing and fundamental geographic datasets have considerably augmented the input variables for SOC estimation models. However, the autocorrelation present among a large number of variables—the so-called "curse of dimensionality"—can compromise the robustness of estimation models. In the context of spatio-temporal big data, the process of feature selection has been identified as a potentially effective approach to reducing data redundancy and noise. This, in turn, has the potential to enhance the reliability of spatial SOC predictions. However, current remote sensing studies on SOC estimation predominantly characterize natural factors while failing to quantitatively represent anthropogenic disturbances-particularly ecological restoration measures-thereby increasing estimation uncertainty. Concurrently, the presence of highly heterogeneous land surfaces, in conjunction with intensive human interventions, has been shown to engender a state of heightened uncertainty in the selection of optimal model factors. This, in turn, has the effect of compromising the accuracy of estimation.

1.3 Influence of Model Algorithms and Data Spatial Resolution on the Estimation Accuracy of SOC

In the aforementioned studies estimating surface SOC content based on environmental variables or remote sensing data, machine learning has been extensively applied, such as Support Vector Machines (SVM), Random Forests (RF) (Zhou et al., 2020), Extreme Gradient Boosting (XGBoost) (Chen et al., 2022) and Cubist. Zhou et al. (2020) enhanced the precision of total soil nitrogen prediction by implementing a weighted average of two machine-learning models, thereby reducing the root mean square error by 2.5–4%. Taghizadeh-Mehrjardi et al. (2020) conducted a comparative analysis of the accuracy of six machine-learning approaches in SOC inversion, determining that Deep Learning Neural Networks (DNN), RF, and XGBoost exhibited superior performance in comparison to SVM, ANN (Artificial Neural Network), and Cubist. In addition, the stacking ensemble learning model constructed reduced the root mean square error by 5–28% across different soil layers in comparison with the six individual models. This enhancement is attributable to the implementation of a stacking integration technique, which integrates the outputs from multiple machine-learning models within a unified framework, thereby ensuring optimal robustness and generalization accuracy. Despite the capacity of machine learning to address complex non-linear problems, the internal logic of such systems remains challenging to interpret. Research has proposed a combination of machine-learning's predictive power with the interpretability of physical models through explainable machine learning (Esterhuizen et al., 2022), alongside collaborative feature selection and sparse counterfactual construction to develop interpretable neural network models (Barraza et al., 2024). To a certain extent, process model parameters can provide interpretability factors for machine-learning model construction.

The spatial resolution of model input variables has been demonstrated to have a significant impact on the accuracy of SOC content estimation (Liu et al., 2020). This is due to the fact that spatial resolution influences the surface information captured by satellite sensors at the pixel level, thereby affecting inversion precision. Castaldi et al. (2019) conducted a comparative analysis of the impact of different spatial resolutions (EnMAP with 30 m resolution and Sentinel-2 with 10 resolution) on the accuracy of soil organic carbon (SOC) prediction, concluding that EnMAP data yielded lower prediction precision. Guo et al. (2019) resampled hyperspectral imagery to 10, 20, and 30 m resolutions, results showed that the estimation accuracy of SOC decreased with the decline of spatial resolution. Spatial downscaling is an effective method of addressing insufficient spatial distribution, with statistical downscaling and data fusion downscaling techniques being widely applied to parameters such as soil moisture, surface temperature, and climate variables (Abowarda et al., 2021; Zhao et al., 2017). The team's High-Accuracy Surface Modeling (HASM) technique has been demonstrated to offer significant advantages in spatial downscaling and data fusion (Yue et al., 2020), with the absolute error of CMIP5 annual mean temperatures in the Beijing-Tianjin-Hebei region reduced by 17% through HASM-DS downscaling.

1.4 Remote Sensing Retrieval of Vegetation Biomass

Vegetation biomass is defined as the total amount of living organic matter within a given plant community at a specific point in time and space. The traditional biomass estimation method is field investigation, but this involves time-consuming, labor-intensive, and highly destructive survey techniques, rendering it suitable only for small-scale biomass assessments. Conducting large-scale vegetation biomass surveys is an exceedingly difficult task. The advent of remote sensing technology has precipitated the progressive implementation of remote sensing inversion in the estimation of biomass. The inherent capacity of optical remote sensing to monitor surface features intuitively led to its primary application in biomass estimation. Depending on the spectral bands utilize, the research encompasses both vegetation biomass estimation based on thermal infrared remote sensing and vegetation biomass estimation based on Synthetic Aperture Radar (SAR).

The present study explores the current state of research in the estimation of aboveground biomass, with a particular focus on the assessment of grassland and forest biomass. Justice and Hiernaux (1986) employed AVHRR data to estimate grassland biomass in the Sahel region of Niger. Subsequently, Wakiyama et al. (1999) utilized LandsatTM remote sensing data to estimate grassland productivity in Japan's Kyushu region. Darvishzadeh et al. (2008) estimated Mediterranean pasture biomass using hyperspectral data. In 2016, Jansen et al. monitored the entire growing season of grassland vegetation in Oregon using Landsat ETM + data (Jansen et al., 2016). Zhou et al. (2021) utilized machine-learning and high-precision surface modeling to achieve remote sensing inversion of aboveground biomass in the Three-rivers source region grasslands. The incorporation of data fusion resulted in a substantial enhancement in simulation accuracy. In 2022, some researchers proposed a novel physics-based aboveground biomass inversion model by inputting leaf area index (LAI) and dry matter mass into the PROSAL model (Xie et al., 2022). Furthermore, research integrating PROSAL and SVR models has been conducted to estimate aboveground biomass (AGB) on China's Ruoergai Plateau, thereby reducing reliance on field sampling points (Wang et al., 2024). The estimation of biomass is significantly influenced by the model employed, with optimal models exhibiting variation across different study areas. These models can be categorize into three broad types: parametric models, process models, and non-parametric estimation models.

It is evident that the field of parametric models is primarily characterized by two distinct classifications: linear parametric models and nonlinear parametric models. Among these, linear regression models are relatively straightforward and facilitate calculation of model parameters; non-linear regression models, being more complex, are typically converted into linear regression models for computational convenience. The primary function of parameter models is to investigate the mathematical relationships between AGB and environmental variables. Luo et al. (2017) employed laser radar and hyperspectral data, utilizing partial least squares (PLS) regression models to estimate the study area's belowground biomass (BGB), AGB, and total forest biomass.

Process models, also referred to as mechanistic models, are characterized by their intricate nature, which distinguishes them from the more straightforward processes associated with parametric models. Mechanistic models are typically utilized to simulate AGB in grasslands and forests. By delineating the various internal processes within these ecosystems, they simulate the relationships between biomass and closely related indicators (such as net primary productivity and leaf area index), thereby indirectly reflecting changes in biomass. The dynamics of these systems are often intricate, given the involvement of numerous parameters, including vegetation growth conditions, climatic variations, topography, and light exposure. It is evident that the procurement of specific parameters continues to present significant challenges, thereby impeding the progression of process models to a considerable extent.

In comparison with parametric processes and process models, non-parametric estimation models do not make any assumptions regarding the underlying function and do not necessitate the input of different types of parameters. Instead, they seek the optimal fit for data points by leveraging the inherent properties and structure of the data itself. Given the susceptibility of AGB to influences such as geographical location, climatic precipitation, and anthropogenic factors, non-parametric estimation models offer greater potential for forest biomass estimation, thereby enhancing both the accuracy and efficiency of biomass assessment. Common non-parametric estimation models include RF, SVM, and XGBoost. Hojo et al. (2023) conducted a comparative analysis of the predictive capabilities of three machine-learning models and one parametric model in forest biomass estimation. The findings of the study demonstrated that machine-learning models outperformed the parametric model in all instances, accurately estimating both low and high forest biomass values while effectively mitigating spectral saturation issues.

1.5 Research Progress on HASM Data Fusion and Its Application in Vegetation and SOC Estimation

The High-Accuracy Surface Modeling (HASM) research program was initiated in the late 1980s. The integration of differential geometry with optimization control theory proved to be a successful approach in resolving the long-standing issues with error and multi-scale problems that had persisted in the field of surface modeling for a period of fifty years. This approach facilitates the seamless integration of micro-process information with macro-pattern information, operating as both a reinforcement learning method and an integrated learning algorithm that represents the optimization process. Its earliest application was in the modeling of cirque morphology (Yue & Ai, 1990). Subsequently, Ecological Surface Modeling was proposed on the basis of high-precision surface modeling techniques and research outcomes in Earth surface system modeling (Yue et al., 2020). This theorem and its related corollaries have been incorporated into the United Nations IPBES guidance document on scenario analysis and modeling methodologies, and are referenced in the Global Assessment Report. In order to enhance the computational speed of HASM, a multigrid method, a modified conjugate gradient method, and an error correction algorithm were established, thus laying the foundation for high-performance computing (Yue & Wang, 2010). In recent years, the global surge in spatio-temporal big data has driven demands for higher temporal and spatial resolution in Earth surface system numerical simulations, alongside three-dimensional visualization. This has posed challenges to parallel computing and operational efficiency. In order to address this challenge, HASM has integrated the quantum computing HHL (Harrow-Hassidim-Lloyd) algorithm (Harrow et al., 2009) with HASM, proposing the HASM-HHL machine-learning algorithm. This methodology has been employed in the construction of a digital terrain model of Wugong Mountain (Yue et al., 2022).

The HASM methodology has been successfully applied to a variety of geographical processes, including spatial interpolation of surface elements across various spatial scales, spatial downscaling and upscaling, data fusion, and model-data assimilation. Examples of such applications include the construction of digital terrain models (Yue et al., 2007), the analysis of climate change (Zhao & Yue, 2014), the fusion of CO_2 concentration data (Yue et al., 2016a, 2016b), the estimation of vegetation carbon stocks (Yue et al., 2016a, 2016b), the modeling of vegetation biomass (Zhou et al., 2021), and the simulation of soil properties. Furthermore, the HASM method has been demonstrated to enhance the downscaling accuracy of future annual mean temperature scenarios by a minimum of 16%, while simultaneously improving the model-data assimilation precision for carbon stocks by 40% (Yue et al., 2020). In the field of soil property modeling, Li et al. (2010) employed HASM to simulate surface SOC content based on topography, climate, and vegetation data, and some other research utilized HASM for spatial prediction of soil properties using qualitative variables of soil and vegetation types (Shi et al., 2011). The findings of both studies demonstrated that the HASM model exhibited superior performance in terms of accuracy when compared with the Kriging interpolation method. This is due to the fact that the HASM model integrates point and area data, thereby yielding a higher level of accuracy than that achievable with either data source in isolation (Yue et al., 2016a, 2016b), while simultaneously capturing the spatial heterogeneity of the simulated object. The objective of this study is to employ the HASM method to achieve efficient fusion of ground-truth measurements and model-simulated surface SOC data.

1.6 Ecological and Environmental Changes on the Tibetan Plateau

The Tibetan Plateau, with an average elevation exceeding 4000 m, is considered the world's highest natural geographical unit in terms of mean altitude. Furthermore, it functions as a pivotal ecological security barrier for both China and Asia. Since the 1980s, global warming has manifested more pronouncedly in the Tibetan Plateau region, with its warming rate approximately double the global average. In this context, it is important to consider how the carbon sink function and ecological environment of the Tibetan Plateau's ecosystems respond to climate change.

Many research based on various methods, such as the remote sensing observations, ecosystem carbon cycle models, and atmospheric inversion models, all found a persistent identification of the Tibetan Plateau ecosystem as a carbon sink. This ecosystem has been found to account for between 10 and 18% of China's terrestrial carbon sink. And the increased of vegetation productivity is the primary carbon sink mechanism on the Tibetan Plateau (Piao et al., 2009). The most significant component of the Tibetan

Plateau's ecosystem carbon sink is constituted by soil carbon pools. Large-scale, repeated soil surveys indicate that over the past decade, the surface soil carbon pool (0–30 cm) in the plateau's alpine grasslands has accumulated at an average rate of 28 g C/m^2/a. In summary, the evidence from both flux observations and field soil survey data supports the conclusion that the Tibetan Plateau has functioned as a carbon sink in recent years. It has been demonstrated that enhanced vegetation growth has led to an improvement in soil water retention capacity. However, given the substantial size of the plateau, the significant spatial heterogeneity of surface elements, and the variations in survey methodologies, considerable uncertainties persist in estimating the magnitude of ecosystem carbon sinks, particularly soil carbon sinks, on the Tibetan Plateau.

It is well know that changes in the structure and function of grassland ecosystems on the Tibetan Plateau are the result of a combination of climate change and human activities (Chen et al., 2020; Zhang et al., 2015a, 2015b). Climate change has been demonstrated to alter temperature, precipitation, and light conditions in ways that may favor or hinder plant growth (Zhang et al., 2014), and to accelerate carbon cycling rates within regional terrestrial ecosystems (Zhang et al., 2015a, 2015b). Concurrently, human activities exert direct or indirect influences on ecosystem structure and function. For instance, grassland management practices affect vegetation distribution patterns and productivity Nevertheless (Zhang et al., 2017), the relative contributions of climate and human activities to grassland productivity and ecosystem degradation on the Tibetan Plateau remain a matter of debate. The prevailing opinion among researchers in this field is that the primary cause of grassland degradation is climate-related, particularly variations in temperature and precipitation (Zhang et al., 2017). However, other studies have suggested that human activities, such as overgrazing and land-use changes, are the dominant drivers of degradation (Lu et al., 2017). Nevertheless, over the past two decades, both climatic change and human disturbance on the plateau have intensified, resulting in increased spatial heterogeneity. Concurrently, substantial investments in ecological engineering projects—such as the Grassland Restoration Programme and the Three Rivers Source Protection Initiative—have contributed to enhanced grassland productivity, rendering ecosystem dynamics increasingly complex (Du et al., 2004). Consequently, the quantitative assessment of the impacts of climate change and human activities on ecosystem changes in the Tibetan Plateau holds significant guiding importance for the implementation of ecosystem conservation measures (Zhang et al., 2019).

References

Abowarda, A. S., Bai, L., Zhang, C., Long, D., Li, X., Huang, Q., & Sun, Z. (2021). Generating surface soil moisture at 30 m spatial resolution using both data fusion and machine learning toward better water resources management at the field scale. *Remote Sensing of Environment, 255*, 112301.

Arora, V. K., Boer, G. J., Friedlingstein, P., Eby, M., Jones, C. D., Christian, J. R., Bonan, G., Bopp, L., Brovkin, V., Cadule, P., Hajima, T., Ilyina, T., Lindsay, K., Tjiputra, J. F., & Wu, T. (2013). Carbon-concentration and carbon-climate feedbacks in CMIP5 earth system models. *Journal of Climate, 26*, 5289–5314.

Ballabio, C., Fava, F., & Rosenmund, A. (2012). A plant ecology approach to digital soil mapping, improving the prediction of soil organic carbon content in alpine grasslands. *Geoderma, 187*, 102–116.

Bangroo, S. A., Najar, G. R., Achin, E., & Truong, P. N. (2020). Application of predictor variables in spatial quantification of soil organic carbon and total nitrogen using regression kriging in the North Kashmir forest Himalayas. *Catena, 193*, 104632.

Barraza, J. F., Droguett, E. L., & Martins, M. R. (2024). FS-SCF network: Neural network interpretability based on counterfactual generation and feature selection for fault diagnosis. *Expert Systems with Applications, 237*, 121670.

Beatriz, L., Parras-Alcantara, L., & Brevik, E. C. (2016). Impact of topographic aspect and vegetation (native and reforested areas) on soil organic carbon and nitrogen budgets in Mediterranean natural areas. *Science of the Total Environment, 544*, 963–970.

Behrens, T., Schmidt, K., Ramirez-Lopez, L., Gallant, J., Zhu, A. X., & Scholten, T. (2014). Hyperscale digital soil mapping and soil formation analysis. *Geoderma, 213*, 578–588.

Biswas, A., Cresswell, H. P., Chau, H. W., Viscarra Rossel, R. A., & Si, B. C. (2013). Separating scale-specific soil spatial variability: A comparison of multi-resolution analysis and empirical mode decomposition. *Geoderma, 209*, 57–64.

Bradford, M. A., Wieder, W. R., Bonan, G. B., Fierer, N., Raymond, P. A., & Crowther, T. W. (2016). Managing uncertainty in soil carbon feedbacks to climate change. *Nature Climate Change, 6*, 751–758.

Calzolari, C., Ungaro, F., & Vacca, A. (2021). Effectiveness of a soil mapping geomatic approach to predict the spatial distribution of soil types and their properties. *Catena, 196*, 104818.

Castaldi, F., Chabrillat, S., & Van Wesemael, B. (2019). Sampling strategies for soil property mapping using multispectral sentinel-2 and hyperspectral EnMAP satellite data. *Remote Sensing, 11*, 1–19.

Chaopricha, N. T., & Marín-Spiotta, E. (2014). Soil burial contributes to deep soil organic carbon storage. *Soil Biology & Biochemistry, 69*, 251–264.

Chen, H., Ju, P. J., Zhang, J., Wang, Y. Y., Zhu, Q. A., Yan, L., Kang, X. M., He, Y. X., Zeng, Y., Hao, Y. B., & Wang, Y. F. (2020). Attribution analyses of changes in alpine grasslands on the Qinghai-Tibetan Plateau. *Chinese Science Bulletin-Chinese, 65*, 2406–2418.

Chen, Y., Ma, L. X., Yu, D. S., Zhang, H. D., Feng, K. Y., Wang, X., & Song, J. (2022). Comparison of feature selection methods for mapping soil organic matter in subtropical restored forests. *Ecological Indicators, 135*, 108545.

Chen, Z. D., Ti, J. S., & Chen, F. (2017). Soil aggregates response to tillage and residue management in a double paddy rice soil of the Southern China. *Nutrient Cycling in Agroecosystems, 109*, 103–114.

Cheng, Y., Cai, Z. C., Chang, S. X., Wang, J., & Zhang, J. B. (2013). Effects of soil pH and salt on N_2O production in adjacent forest and grassland soils in central Alberta, Canada. *Journal of Soils and Sediments, 13*, 863–868.

Darvishzadeh, R., Skidmore, A., Schlerf, M., Atzberger, C., Corsi, F., & Cho, M. (2008). LAI and chlorophyll estimation for a heterogeneous grassland using hyperspectral measurements. *Isprs Journal of Photogrammetry and Remote Sensing, 63*, 409–426.

Demattê, J., Sayo, V. M., Rizzo, R., & Fongaro, C. T. (2017). Soil class and attribute dynamics and their relationship with natural vegetation based on satellite remote sensing. *Geoderma, 302*, 39–51.

Du, M., Kawashima, S., Yonemura, S., Zhang, X., & Chen, S. (2004). Mutual influence between human activities and climate change in the Tibetan Plateau during recent years. *Global and Planetary Change, 41*, 241–249.

Esterhuizen, J. A., Goldsmith, B. R., & Linic, S. (2022). Interpretable machine learning for knowledge generation in heterogeneous catalysis. *Nature Catalysis, 5*, 175–184.
Filippi, P., Cattle, S. R., Bishop, T. F. A., Odeh, I. O. A., & Pringle, M. J. (2018). Digital soil monitoring of top- and sub-soil pH with bivariate linear mixed models. *Geoderma, 322*, 149–162.
Fu, X. L., Shao, M. A., Wei, X. R., & Horton, R. (2010). Soil organic carbon and total nitrogen as affected by vegetation types in Northern Loess Plateau of China. *Geoderma, 155*, 31–35.
Garnier, J., Billen, G., Tournebize, J., Barre, P., Mary, B., & Baudin, F. (2022). Storage or loss of soil active carbon in cropland soils: The effect of agricultural practices and hydrology. *Geoderma, 407*, 115538.
Gholizadeh, A., Žižala, D., Saberioon, M., & Borůvka, L. (2018). Soil organic carbon and texture retrieving and mapping using proximal, airborne and Sentinel-2 spectral imaging. *Remote Sensing of Environment, 218*, 89–103.
Guo, L., Shi, T. Z., Linderman, M., Chen, Y. Y., & Fu, P. (2019). Exploring the influence of spatial resolution on the digital mapping of soil organic carbon by airborne hyperspectral VNIR imaging. *Remote Sensing, 11*, 10–3390.
Harrow, A. W., Hassidim, A., & Lloyd, S. (2009). Quantum algorithm for linear systems of equations. *Physical Review Letters, 103*, 150502.
Hengl, T., De Jesus, J. M., Heuvelink, G. B. M., Maria, R. G., Milan, K., Aleksandar, B., Wei, S., Wright, M. N., Xiaoyuan, G., & Bernhard, B. M. (2017). Soil Grids 250m: Global gridded soil information based on machine learning. *PLoS ONE, 12*, e169748.
Hojo, A., Avtar, R., Nakaji, T., Tadono, T., & Takagi, K. (2023). Modeling forest above-ground biomass using freely available satellite and multisource datasets. *Ecological Informatics, 74*, 101973.
Hong, S. B., Yin, G. D., Piao, S. L., Dybzinski, R., Cong, N., Li, X. Y., Wang, K., Peñuelas, J., Zeng, H., & Chen, A. P. (2020). Divergent responses of soil organic carbon to afforestation. *Nature Sustainability, 3*, 694.
Jansen, V. S., Kolden, C. A., Taylor, R. V., & Newingham, B. A. (2016). Quantifying livestock effects on bunchgrass vegetation with Landsat ETM plus data across a single growing season. *International Journal of Remote Sensing, 37*, 150–175.
Jobbágy, E. G., & Jackson, R. B. (2000). The vertical distribution of soil organic carbon and its relation to climate and vegetation. *Ecological Applications, 10*, 423–436.
Justice, C. O., & Hiernaux, P. (1986). Monitoring the grasslands of the Sahel using NOAA AVHRR data—Niger 1983. *International Journal of Remote Sensing, 7*, 1475–1497.
Karunaratne, S. B., Bishop, T., Baldock, J. A., & Odeh, I. (2014). Catchment scale mapping of measureable soil organic carbon fractions. *Geoderma, 219*, 14–23.
Lal, R. (2004). Soil carbon sequestration impacts on global climate change and food security. *Science, 304*, 1623–1627.
Lamichhane, S., Kumar, L., & Wilson, B. (2019). Digital soil mapping algorithms and covariates for soil organic carbon mapping and their implications: A review. *Geoderma, 352*, 395–413.
Lan, J. C. (2021). Responses of soil organic carbon components and their sensitivity to karst rocky desertification control measures in Southwest China. *Journal of Soils and Sediments, 21*, 1–12.
Li, J. Y., Zhang, D. Y., & Liu, M. (2020). Factors controlling the spatial distribution of soil organic carbon in Daxing'anling Mountain. *Scientific Reports, 10*, 1–8.
Li, Q. Q., Yue, T. X., Fan, Z. M., & Du, Z. P. (2010). Study on method for spatial simulation of topsoil SOM at national scale in China. *Journal of Natural Resources, 25*, 1385–1399.
Li, Z. W., Liu, L., Nie, X. D., Chang, X. F., Liu, C., & Xiao, H. B. (2016). Modeling soil organic carbon loss in relation to flow velocity and slope on the loess Plateau of China. *Soil Science Society of America Journal, 80*, 1341–1351.
Liu, F., Wu, H. Y., Zhao, Y. G., Li, D. C., Yang, J. L., Song, X. D., Shi, Z., Zhu, A. X., & Zhang, G. L. (2022). Mapping high resolution national soil information grids of China. *Science Bulletin, 67*, 328–340.
Liu, M. J., Xu, Z. K., Gao, Y. B., & Yang, J. (2020). Estimating soil organic matter based on machine learning under sparse sample. *Journal of Geo-Information Science, 22*, 1799–1813.

Liu, R., Wang, M. E., & Chen, W. P. (2018). The influence of urbanization on organic carbon sequestration and cycling in soils of Beijing. *Landscape and Urban Planning, 169*, 241–249.
Liu, Y. N., Min, X., Zhang, X. L., Yu, Z. D., & Kong, F. L. (2019). Carbon storage distribution characteristics of wetlands in China and its influencing factors. *Chinese Journal of Applied Ecology, 30*, 2481–2489.
Lu, X. Y., Kelsey, K. C., Yan, Y., Sun, J., Wang, X. D., Cheng, G. W., & Neff, J. C. (2017). Effects of grazing on ecosystem structure and function of alpine grasslands in Qinghai-Tibetan Plateau: a synthesis. *Ecosphere, 8*, e01656.
Luo, S. Z., Wang, C., Xi, X. H., Pan, F. F., Peng, D. L., Zou, J., Nie, S., & Qin, H. M. (2017). Fusion of airborne LiDAR data and hyperspectral imagery for aboveground and belowground forest biomass estimation. *Ecological Indicators, 73*, 378–387.
Mandal, D., Kumar, V., Ratha, D., Dey, S., & Rao, Y. S. (2020). Dual polarimetric radar vegetation index for crop growth monitoring using sentinel-1 SAR data. *Remote Sensing of Environment, 247*, 1–17.
Meng, X. T., Bao, Y. L., Liu, J. G., Liu, H. J., & Kong, F. C. (2020). Regional soil organic carbon prediction model based on a discrete wavelet analysis of hyperspectral satellite data. *International Journal of Applied Earth Observation and Geoinformation, 89*, 1–15.
Mishra, U., Lal, R., Liu, D., & Meirvenne, M. V. (2010). Predicting the spatial variation of the soil organic carbon pool at a regional scale. *Soil Science Society of America Journal, 74*, 1–9.
Piao, S., Fang, J., Ciais, P., Peylin, P., Huang, Y., Sitch, S., & Wang, T. (2009). The carbon balance of terrestrial ecosystems in China. *Nature, 458*, 1009–1013.
Poggio, L., & Gimona, A. (2017). Assimilation of optical and radar remote sensing data in 3D mapping of soil properties over large areas. *Science of the Total Environment, 579*, 1094–1110.
Shi, W. J., Liu, J. Y., Du, Z. P., & Yue, T. X. (2011). High accuracy surface modeling of soil properties based on geographic information. *Acta Geographica Sinica, 66*, 1574–1581.
Sinoga, J., Pariente, S., Diaz, A. R., & Murillo, J. (2012). Variability of relationships between soil organic carbon and some soil properties in Mediterranean rangelands under different climatic conditions (South of Spain). *Catena, 94*, 17–25.
Sun, Z. X., Li, Y., Zhao, Y. Z., Huang, Y. F., Guo, X. L., & Cao, M. (2019). Analysisi on spatial distribution characteristics and driving forces of soil organic carbon density in dry farming region. *Transactions of the Chinese Society for Agricultural Machinery, 50*, 255–262.
Sutfin, N. A., Wohl, E. E., & Dwire, K. A. (2016). Banking carbon: A review of organic carbon storage and physical factors influencing retention in floodplains and riparian ecosystems. *Earth Surface Processes and Landforms, 41*, 38–60.
Taghizadeh-Mehrjardi, R., Schmidt, K., Chakan, A. A., Rentschler, T., & Scholten, T. (2020). Improving the spatial prediction of soil organic carbon content in two contrasting climatic regions by stacking machine learning models and rescanning covariate space. *Remote Sensing, 12*, 1095.
Wagner, S., Cattle, S. R., & Scholten, T. (2007). Soil-aggregate formation as influenced by clay content and organic-matter amendment. *Journal of Plant Nutrition and Soil Science, 170*, 173–180.
Wakiyama, Y., Inoue, T., & Koyama, N. (1999). Prediction of grass production by using LANDSAT TM data in the Kyushu region. *Japanese Journal of Crop Science, 68*, 301–305.
Wang, Z. F., He, L., He, Z. W., Wang, X. M., Li, L. L., Kang, G. C., Bai, W. Q., Chen, X., Zhao, Y., & Xiao, Y. X. (2024). Integrating the PROSAIL and SVR models to facilitate the inversion of grassland aboveground biomass: A case study of Zoigê Plateau, China. *Remote Sensing, 16*, 1117.
Wiaux, F., Cornelis, J. T., Cao, W., Vanclooster, M., & Van Oost, K. (2014). Combined effect of geomorphic and pedogenic processes on the distribution of soil organic carbon quality along an eroding hillslope on loess soil. *Geoderma, 216*, 36–47.
Xie, J. L., Wang, C. J., Ma, D. J., Chen, R., Xie, Q. Y., Xu, B. D., Zhao, W., & Yin, G. F. (2022). Generating spatiotemporally continuous grassland aboveground biomass on the Tibetan Plateau

through PROSAIL model inversion on Google earth engine. *IEEE Transactions on Geoscience & Remote Sensing, 60*, 1–10.
Xie, X. L., Sun, B., Zhou, H. Z., & Li, Z. P. (2004). Soil carbon stocks and their influencing factors under native vegetations in China. *Acta Pedologica Sinica, 41*, 687–699.
Yue, T. X., Du, Z. P., Song, D. J., & Gong, Y. (2007). A new method of surface modeling and its application to DEM construction. *Geomorphology, 91*, 161–172.
Yue, T. X., Liu, Y., Du, Z. P., Wilson, J., Zhao, D. Y., Wang, Y., Zhao, N., Shi, W. J., Fan, Z. M., Zhao, X. M., Zhang, Q., Huang, H. S., Wu, Q. Y., Zhou, W., Jiao, Y. M., Xu, Z., Li, S. B., Yang, Y., & Fu, B. J. (2022). Quantum machine learning of eco-environmental surfaces. *Science Bulletin, 67*, 1031–1033.
Yue, T. X., Wang, Y. F., Du, Z. P., Zhao, M. W., Zhang, L. L., Zhao, N., Lu, M., Larocque, C. R., & Wilson, J. P. (2016a). Analysing the uncertainty of estimating forest carbon stocks in China. *Biogeosciences, 13*, 3991–4004.
Yue, T. X., Zhao, N., Fan, Z. M., Li, J., Chen, C. F., Lu, Y. M., Wang, C. L., Xu, B., & Wilson, J. (2016b). CMIP5 downscaling and its uncertainty in China. *Global & Planetary Change, 146*, 30–37.
Yue, T. X., Zhao, N., Liu, Y., Wang, Y. F., Zhang, B., Du, Z. P., Wei, Z., & Wilson, J. P. (2020). A fundamental theorem for eco-environmental surface modelling and its applications. *Science China Earth Science, 63*, 1092–1112.
Yue, T. X., & Ai, N. S. (1990). The mathematic model of cirque morphology. *Journal of Glaciology and Geocryology, 12*, 227–234.
Yue, T. X., & Wang, S. H. (2010). Adjustment computation of HASM: A high-accuracy and high-speed method. *International Journal of Geographical Information Science, 24*, 1725–1743.
Zhang, M. L., Lal, R., Zhao, Y. Y., Jiang, W. L., & Chen, Q. G. (2017). Spatial and temporal variability in the net primary production of grassland in China and its relation to climate factors. *Plant Ecology, 218*, 1117–1133.
Zhang, X., Li, X., Ji, X. D., Zhang, Z. Q., Zhang, H. S., Zha, T. G., & Jiang, L. (2021a). Elevation and total nitrogen are the critical factors that control the spatial distribution of soil organic carbon content in the shrubland on the Bashang Plateau. *China. Catena, 204*, 105415.
Zhang, X. Z., Yang, Y. P., Piao, S. L., Bao, W. K., Wang, S. P., Wang, G. X., Sun, H., Luo, T. X., Zhang, Y. J., Shi, P. L., Liang, E. Y., Shen, M. G., Wang, J. S., Gao, Q. Z., Zhang, Y. L., & Ouyang, H. (2015a). Ecological change on the Tibetan Plateau. *Chinese Science Bulletin, 60*, 3048–3056.
Zhang, Y. J., Zhu, Y. X., Li, J. X., & Chen, Y. (2019). Current status and future directions of the Tibetan Plateau ecosystem research. *Science Bulletin, 64*, 428–430.
Zhang, Y. L., Hu, Z. J., Qi, W., Xue, W., Bai, W. Q., Li, L. H., Ding, M. J., Liu, L. S., Wang, Z. F., & Zheng, D. (2015b). Assessment of protection effectiveness of nature reserves on the Tibetan Plateau based on net primary priduction and the large-sample-comparison method. *Acta Geographical Sinica, 70*, 1027–1040.
Zhang, Y. L., Qi, W., Zhou, C. P., Ding, M. J., Liu, L. S., Gao, J. G., Bai, W. Q., Wang, Z. F., & Zheng, D. (2014). Spatial and temporal variability in the net primary production of alpine grassland on the Tibetan Plateau since 1982. *Journal of Geographical Sciences, 24*, 269–287.
Zhang, Y. Q., Ai, J. J., Sun, Q. W., Li, Z. C., Hou, L. Y., Song, L. G., Tang, G. Y., Li, L., & Shao, G. D. (2021b). Soil organic carbon and total nitrogen stocks as affected by vegetation types and altitude across the mountainous regions in the Yunnan Province, south-western China. *Catena, 196*, 104875.
Zhao, N., & Yue, T. X. (2014). A modification of HASM for interpolating precipitation in China. *Theoretical & Applied Climatology, 116*, 273–285.
Zhao, W., Li, A. N., Jin, H. A., Zhang, Z. J., Bian, J. H., & Yin, G. F. (2017). Performance evaluation of the triangle-based empirical soil moisture relationship models based on Landsat-5 TM data and in situ measurements. *IEEE Transactions on Geoscience and Remote Sensing, 55*, 2632–2645.

Zhao, Y. C., Wang, M. Y., Hu, S. J., Zhang, X. D., Ouyang, Z., Zhang, G. L., Huang, B. A., Zhao, S. W., Wu, J. S., Xie, D. T., Zhu, B., Yu, D. S., Pan, X. Z., Xu, S. X., & Shi, X. Z. (2018). Economics- and policy-driven organic carbon input enhancement dominates soil organic carbon accumulation in Chinese croplands. *Proceedings of the National Academy of Sciences of the United States of America, 115*, 4045–4050.

Zhou, G. S., Wang, Y. H., Jiang, Y. L., & Yang, L. M. (2002). Conversion of terrestrial ecosystems and carbon cycling. *Acta Phytoecologica Sinica, 26*, 250–254.

Zhou, T., Geng, Y. J., Chen, J., Liu, M. M., & Lausch, A. (2020). Mapping soil organic carbon content using multi-source remote sensing variables in the Heihe River Basin in China. *Ecological Indicators, 114*, 106288.

Zhou, W., Li, H., Xie, L., Nie, X., Wang, Z., Du, Z., & Yue, T. (2021). Remote sensing inversion of grassland aboveground biomass based on high accuracy surface modeling. *Ecological Indicators, 121*, 107215.

Chapter 2
Estimation of Biomass Carbon Density Using Multi-source Data Over the Tibetan Plateau

Abstract Biomass carbon plays a crucial role in the global carbon cycle and carbon sinks in terrestrial ecosystems. The Tibetan Plateau is an amplifier of global climate change, and the spatial distribution of biomass in the region responds with high sensitivity to climate change. In addition to climate change, human activities interventions also greatly affect the carbon cycle and carbon balance in the alpine ecosystem of the Tibetan Plateau. In this study, biomass carbon density and spatial distribution on the Tibetan Plateau from 2001 to 2020 was conducted using three machine-learning (ML) algorithms, including random forest (RF), support vector machines (SVM) and extreme gradient boosting (XGBoost). Multi-source environmental variables were comprehensively considered, including topographic factors, climatic factors, biological factors, soil properties and socioeconomic factors. The results showed that: (1) The average aboveground biomass carbon density was 3.864 kg C/m^2, and the average belowground biomass carbon density was 0.918 kg C/m^2. (2) Among the three ML algorithms, the validation accuracy of RF was the highest, and its accuracy for AGB were higher (R^2 = 0.519, RMSE = 1.283 kg C/m^2) than BGB (R^2 = 0.378, RMSE = 0.705 kg C/m^2). (3) Soil pH, precipitation, and livestock volume were important indicators for the prediction of AGB, with importance values of 18.93%, 13.50%, and 7.18%, respectively; latitude, precipitation, and soil organic carbon were important indicators for the prediction of BGB, with importance values of 7.23%, 7.09%, and 6.11%, respectively. (4) Spatially, the distribution pattern of AGB and BGB of Tibetan Plateau was consistent, and the high value was located in the east and low in the west, which was consistent with the distribution pattern of the sampling sites data. In a word, accurate accounting and spatiotemporal dynamics of biomass carbon density can provide important theoretical basis for carbon sink assessment.

Keywords Biomass carbon · Machine learning · Tibetan Plateau · Uncertainty analysis

© The Author(s), under exclusive license to Springer Nature Singapore Pte Ltd. 2026
W. Zhou et al., *Carbon Sink Estimation and Evaluation of Ecological Function Changes on the Tibetan Plateau*, https://doi.org/10.1007/978-981-95-8664-6_2

2.1 Introduction

Biomass, in general, includes both aboveground biomass and belowground biomass. Biomass carbon, as an important component of the carbon pool in terrestrial ecosystems, plays an important role in linking the carbon cycle with the atmospheric carbon pool (Hari & Tyagi, 2022). The carbon pool of terrestrial ecosystems, as an important part of the Earth's carbon pool (Wani et al., 2023; Wu et al., 2020), is able to absorb atmospheric carbon dioxide for carbon sequestration (Keenan & Williams, 2018), which contributes to slowing down the trend of global warming.

Biomass is one of the important indicators for evaluating terrestrial ecosystems, however, the characteristics of biomass growth over time (spatio-temporal heterogeneity) bring great difficulties and high uncertainty to biomass monitoring and modeling (Qin et al., 2021). Overall, biomass estimation methods can be classified into field measurements, remote sensing-based methods and ecological model simulations (Huang et al., 2019; Zhang et al., 2019). Traditional biomass estimation methods, survey methods are time-consuming, labor-intensive and destructive, which are only suitable for biomass estimation in small-scale areas, and it is difficult to carry out large-area vegetation biomass surveys (Lu et al., 2016; Sinha et al., 2015). Although the biomass measured in the field is the most accurate, it is not a method for estimating biomass on a large scale (Kumar & Mutanga, 2017). In recent years, with the development of remote sensing technology, optical remote sensing has been firstly applied to the field of biomass estimation due to its intuitive nature of monitoring surface features (Jansen et al., 2016; Koju et al., 2017; Meng et al., 2020; Nichol & Sarker, 2011). In addition to optical remote sensing, thermal infrared remote sensing data, microwave remote sensing data, and radar remote sensing data can be used for biomass estimation, depending on the wavelength band used (Lu, 2006). Radar remote sensing has gained prominence in aboveground biomass estimation due to its ability to penetrate clouds and its detailed information on vegetation structure (Kumar et al., 2015). Biomass estimation based on multi-source remote sensing data has also been a research hotspot in recent years; for example, the optical remote sensing and radar data was used to carry out biomass estimation with higher precision and accuracy (Pan et al., 2020; Zhang et al., 2022). In addition, ecological process model is also a promising method for biomass dynamic evaluation (Tian et al., 2017). But, the process of change is often complex and some parameters are more difficult to obtain due to the involvement of many parameters such as vegetation growth conditions, climate change, topography and light, which limits the development of process modeling to a certain extent (Hurtt et al., 2010; Soenen et al., 2010).

The Tibetan Plateau is a sensitive area for climate change, in which the vegetation carbon stock is 2.4 Pg, accounting for 23.4% of China's vegetation carbon stock (Han et al., 2022), and the region's vegetation carbon stock plays an important role in the carbon cycle of terrestrial ecosystems (Stockmann et al., 2013; Wei et al., 2021). In recent years, due to the dual impacts of global climate change and human activities, vegetation in the Tibetan Plateau has undergone significant changes, which have greatly affected the carbon cycle and carbon balance of the alpine ecosystem (Ma et al., 2021). In the context of global warming and the impact of human activities activities (e.g., grassland degradation), an in-depth study of vegetation carbon density and its changing law on the Tibetan Plateau will provide a scientific basis for strengthening the carbon sequestration capacity of the alpine ecosystem and ecosystem protection. This study combines topographic, climatic, biological and human activity multi-source data, and uses three ML models, RF, SVM, and XGBoost, to predict biomass carbon density on the Tibetan Plateau. The contents of this study are as follows: (1) Calculate the biomass density of the Tibetan Plateau from 2001 to 2020; (2) Explore the change trend of the biomass carbon density of the Tibetan Plateau from 2001 to 2020; (3) Identify the main drivers affecting the spatial change of biomass carbon density in the Tibetan Plateau. This will provide scientific references for terrestrial carbon sequestration and regional sustainable development in the region.

2.2 Materials and Methods

2.2.1 *Study Area*

The Tibetan Plateau is mainly located in Qinghai Province and Tibet Autonomous Region in the western part of China, with the geographical coordinates of 26°00′–39°47′N, 73°19′–104°47′E (Fig. 2.1). The Tibetan Plateau occupied a total area of about 2.5 million km^2. The Tibetan Plateau is the world's highest terrain of the third pole of the terraces, with an average altitude of 4500 m. The glacial and permafrost are widely distributed, with glaciers on the plateau covering an area of about 47,000 square km^2, accounting for more than 80% of the China's total glacier area, and permafrost covering an area of 1.5 million km^2.

2.2.2 *Measured Biomass Data*

Sampling data of biomass comes from the 2010s Chinese terrestrial ecosystem carbon density dataset published by Xu Li on the China Scientific Data Network (CSDN), with a time frame of 2000 to 2014 (Xu et al., 2019). The source of this dataset consists of two parts: one part is the experimental data of the group, and the other part is the officially published literature data. There are 945 AGB and 764 BGB field sampling data in the study area.

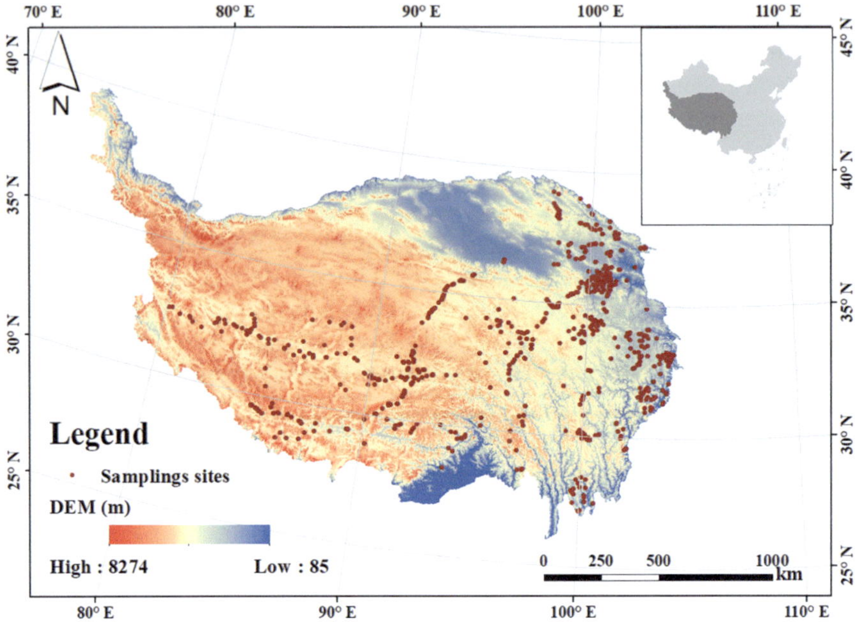

Fig. 2.1 Spatial location of the study area and biomass sampling sites

2.2.3 Environmental Covariate Data

The biomass is affected by many factors. Combined with the relevant literature, we analyzed the topography, climate, soil, biology, and human activities, and selected a total of 21 environmental variables, including topographic factors (elevation, slope, slope direction, topographic suitability index, topographic location index, topographic durability index, longitude, latitude), climatic factors (temperature, precipitation, land-surface temperature data), soil factors (soil clay content, soil silt content, soil sand content, soil organic carbon, soil pH), biological factors (normalized difference vegetation index, augmented vegetation index), and human activities factors (livestock population, population, night lighting).

2.2.4 Methods

Random forests (RF) were proposed by Breiman (2001). The RF model is a relatively new machine-learning model that is mainly applied to address classification and regression problems. It essentially belongs to the integration learning method, which performs regression prediction by combining multiple decision trees, and the biggest benefit is that it can find the best prediction in complex environmental variables due to the non-linear relationship between the variables and multiple independent variables, and fit them together (Biau, 2012; Fawagreh et al., 2014).

The basic idea of support vector machine (SVM) is to use the kernel function to map linearly indistinguishable points in the low-dimensional input space into linearly distinguishable points in the high-dimensional feature space, and to achieve the maximum generalization ability by dividing the hyperplane so that the similarity between similar samples is as large as possible, i.e., the distance of all the points to the classification hyperplane is maximized (Hearst et al., 1998; Tian et al., 2012).

Extreme Gradient Boosting (XGBoost) algorithm is a new Gradient Boosting Decision Tree Algorithm (GBDT) proposed by Chen and Guestrin which is based on augmented decision trees and is capable of handling weakly supervised learning in machine learning for classification and return tasks through a training process. The XGBoost model is designed to reduce the computational cost through simplification and regularization, while ensuring optimal computational efficiency in order to obtain accurate predictions and avoid overfitting. XGBoost outperforms traditional GBDT algorithms by improving the accuracy of obtaining optimal solutions through a second-order Taylor expansion of the loss function (Chen & Guestrin, 2016).

2.3 Results and Analysis

2.3.1 Model Driving Factors Selecting

Firstly, correlation coefficient analysis was analyzed between 21 environmental variables and biomass value of 945 AGB and 764 BGB field sampling points (Fig. 2.2): Among the topographic factors, DEM showed a significant negative correlation ($P < 0.001$) with aboveground biomass (AGB), which is due to the fact that the high-altitude areas of the Tibetan Plateau are covered by snow and ice all year round, which limits the light and heat conditions required for plant growth and slows down the accumulation of biomass; the topographic moisture index (TWI) showed a significant negative correlation ($P < 0.01$) with belowground biomass (BGB), which is an indicator to quantify soil moisture content. Index (TWI) was significantly negatively correlated with belowground biomass (BGB) ($P < 0.01$), and TWI is an index to quantify soil moisture content, thus belowground biomass decreased with the decrease of soil moisture. Among the biological factors, NDVI, EVI, and AGB, BGB were all significantly positively correlated ($P > 0.001$). Among the soil factors, there were positive correlation between soil clay content, soil flour content, soil organic carbon (SOC) and AGB, and BGB ($P > 0.001$); there were negatively correlated between soil sand content, soil pH ($P < 0.001$). Because, with the coarser grained soil texture, the soil capacity was degraded, which was unfavorable to the growth of vegetation, whereas SOC affects soil fertility as well as vegetation productivity and is one of the important influencing factors of biomass accumulation. Among the climatic factors, air temperature, land surface temperature (LST), and AGB showed highly significant positive correlation ($P > 0.001$), and since the northwestern part of the Tibetan Plateau belongs to the alpine region, higher temperature has a promoting effect on vegetation growth in the alpine region. Among the human activities factors, livestock volume (GLW) was significantly negatively correlated with AGB ($P < 0.01$), and the aboveground biomass decreased with the increase of grazing intensity.

In addition, during the modeling process of RF, it was possible to analyze the importance of the environmental variables (Fig. 2.3), in which the variables of aboveground biomass that ranked high in importance were: soil pH, precipitation, livestock, soil clay content, and longitude, with importance values of 18.93%, 13.50%, 7.18%, 6.97%, and 6.51%, respectively; The variables that ranked high in importance for belowground biomass were: latitude, precipitation, soil organic carbon, terrain durability index, and soil sand content, with importance values of 7.23%, 7.09%, 6.11%, 5.92%, and 5.81%, respectively.

Considering the correlation analysis and variable importance selection of relevant factors, the driving factors for model construction were determined. For AGB modeling, totally 17 environmental variables were selected, another four factors were ignored including slope direction, topographic location index, population, and nighttime light index. For BGB modeling, totally 15 environmental variables were selected, another six factors were ignored including elevation, slope direction, topographic location index, population, air temperature, and latitude were ignored.

2.3.2 *Model Accuracy Evaluation*

The above screened variables were used as independent variables and the measured biomass data as dependent variables, and the tenfold cross-validation method was used to estimate aboveground biomass and belowground organisms using three machine-learning models, namely RF, XGBoost, and SVM, respectively, and the accuracy of the models was evaluated using R^2, RSME, and MAE (Table 2.1).

For AGB estimation, RF model achieved the highest estimation accuracy with $R^2 = 0.519$, and the smallest values of RSME and MAE were 1.283 kg C/m^2 and 0.386 kg C/m^2. Respectively, the accuracy of XGBoost and SVM models was lower (Table 2.1). For BGB estimation, RF model also achieved the highest accuracy with $R^2 = 0.378$, and had the smallest values of RSME (0.705 kg C/m^2) and MAE (0.491 kg C/m^2), and SVM model had the lowest accuracy with $R^2 = 0.191$.

Comparing the estimation accuracy of AGB and BGB, AGB achieved higher accuracy than BGB for the three models (Fig. 2.4). Analyzing the reasons, on the one hand, the number of sampling points in the field for BGB is less than that for AGB, on the other hand, the selection of variables also affected.

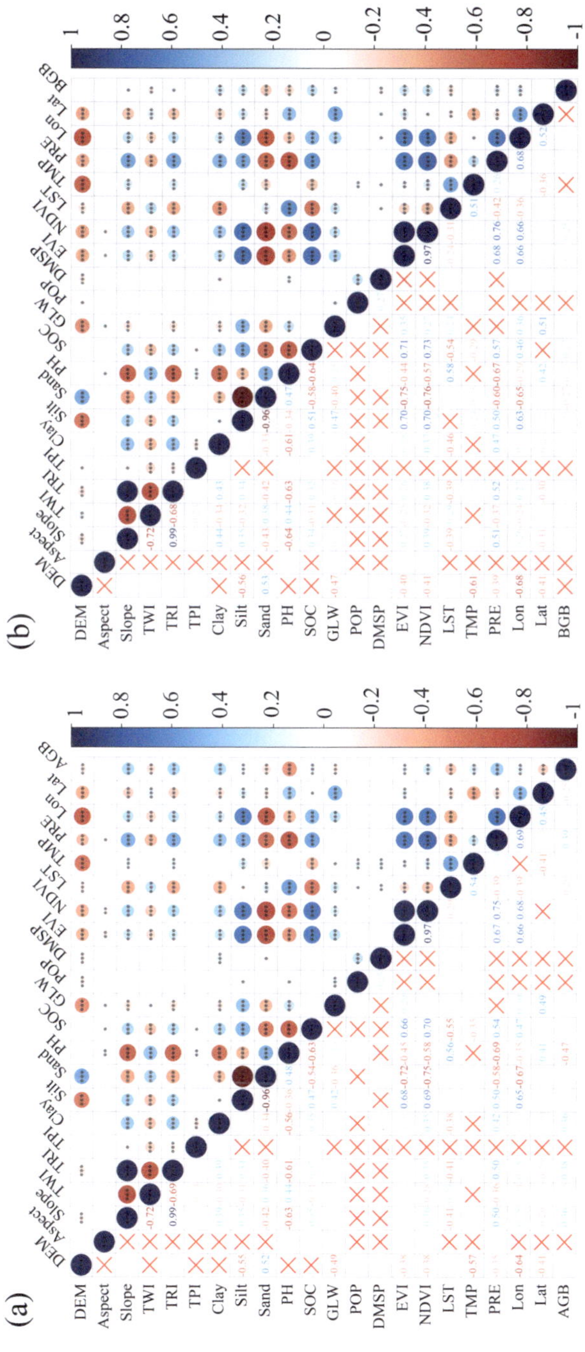

Fig. 2.2 Correlation analysis of environmental covariates of AGB (a), BGB (b)

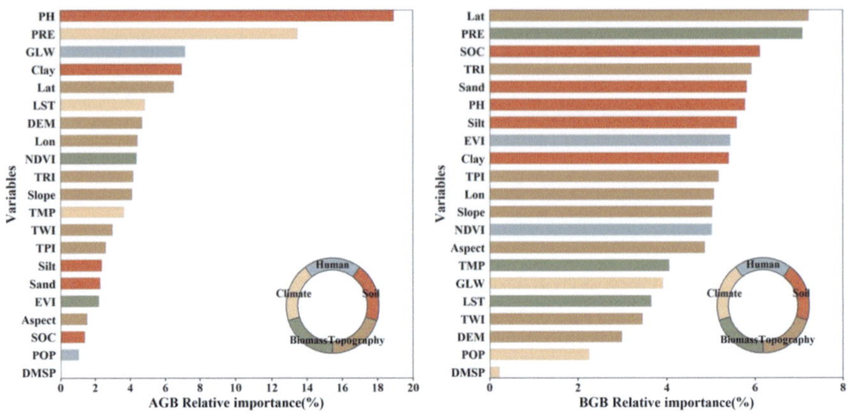

Fig. 2.3 Relative importance of AGB, BGB

Table 2.1 Model prediction accuracy for AGB, BGB

Model	AGB			BGB		
	R^2	RMSE	MAE	R^2	RMSE	MAE
RF	0.519	1.283	0.386	0.378	0.705	0.491
XGBoost	0.392	1.465	0.402	0.310	0.781	0.514
SVM	0.389	1.552	0.451	0.191	0.814	0.542

2.3.3 *Spatial–Temporal Analysis of Biomass Density*

Long time series of AGB and BGB were simulated and predicted from 2001 to 2020 (Fig. 2.5). The predicted values of AGB ranged from 0.009 to 11.656 kg C/m^2, with a mean of 3.864 kg C/m^2, and that of BGB ranged from 0.069 to 3.081 kg C/m^2, with a mean of 0.918 kg C/m^2. The spatial distribution pattern of biomass was similar to the distribution of elevation. The high-elevation areas in the northwestern part of the Tibetan Plateau being covered by snow and ice all year round, with lower values of biomass, because the temperature decreasing with increasing of elevation, is harmful to vegetation growth. The high values of biomass were concentrated in the forested areas in the southern part of the Tibetan Plateau. Specifically, AGB was four times higher than BGB, low AGB values were focused on the northern part of Tibetan Plateau, and low BGB values were focused on the western part of Tibetan Plateau.

For the temporal change trend of AGB and BGB (Fig. 2.6), AGB values increased annually, with a minimum of 3.726 kg C/m^2 in 2001 and a maximum of 3.904 kg C/m^2 in 2020, while the BGB values fluctuated with no obvious increasing or decreasing trend, with a minimum of 0.908 kg C/m^2 in 2004 and a maximum of 0.923 kg C/m^2 in 2005.

For the spatial distribution of biomass carbon density, AGB showed an overall increasing trend from 2001 to 2020, with an increasing trend concentrated in the

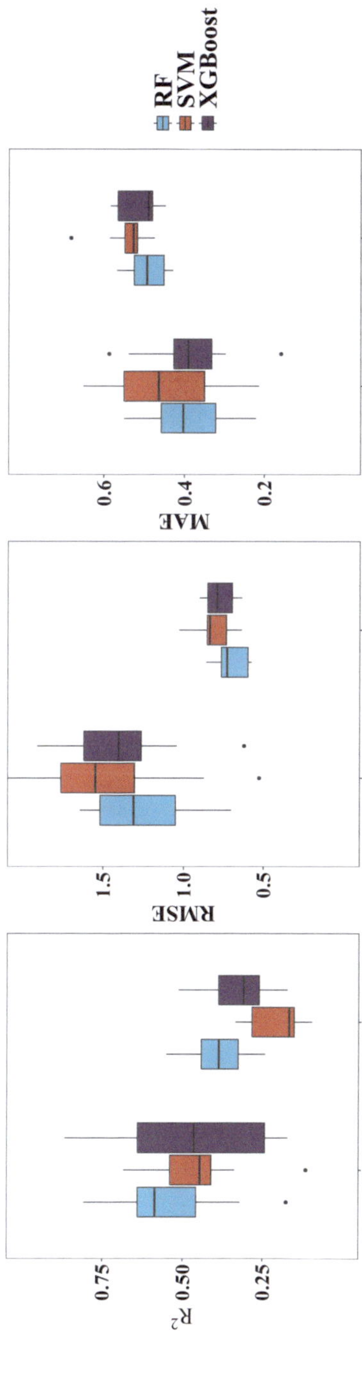

Fig. 2.4 Precision indicators box plot of AGB, BGB

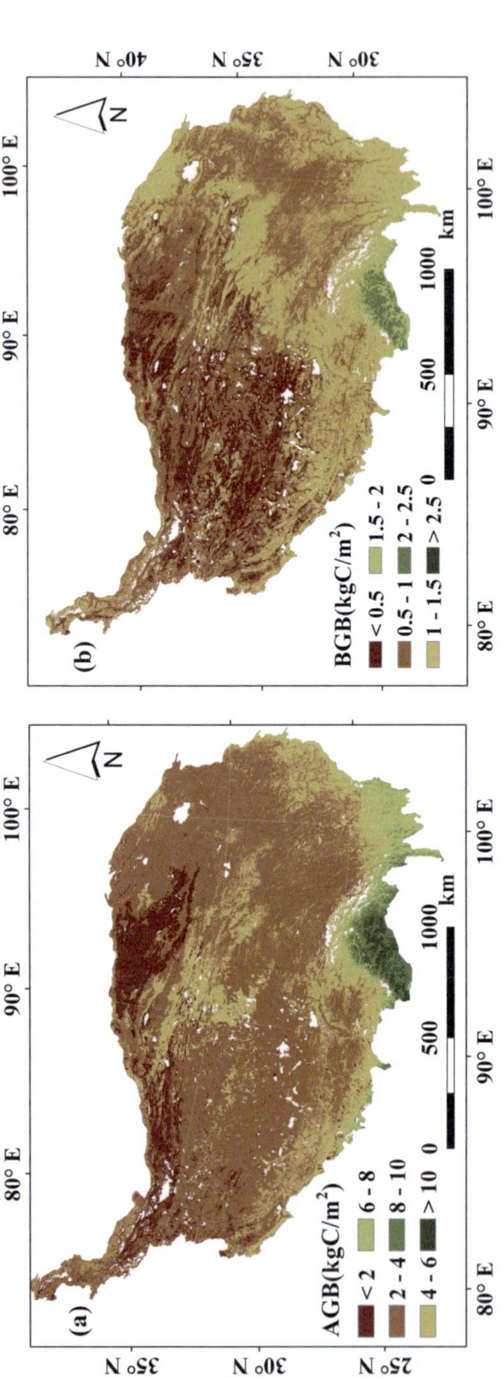

Fig. 2.5 Spatial distribution of biomass carbon density based on RF model, **a** aboveground biomass, **b** belowground biomass

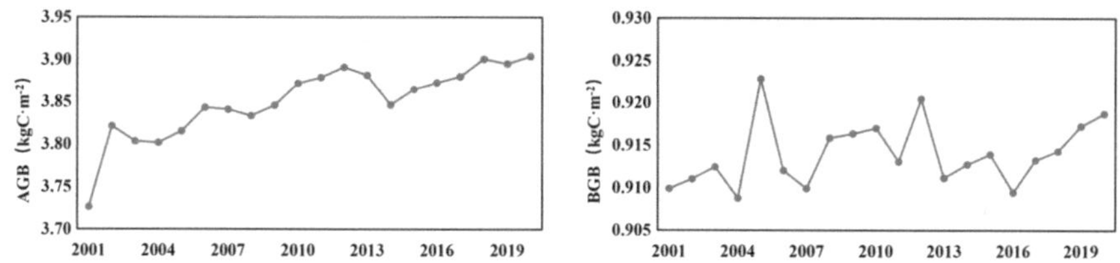

Fig. 2.6 Above- and belowground biomass carbon density time-series variation

northern region and a decreasing trend in the southern region (Fig. 2.7). Among them, the 2005–2010 and 2015–2020 periods are consistent with each other, and the 2015–2020 period shows a more obvious trend of increase and decrease, with a greater trend of change in aboveground biomass carbon density. In contrast, the trend of the 2010–2015 period was the opposite, with a more obvious decrease in the eastern region, and more obvious decreasing and increasing trends in the north and south, respectively.

The increase area in BGB carbon density from 2001 to 2020 was greater than the decrease area in the area, and the trend of change was patchy, with no obvious trend of change in the northern and western regions (Fig. 2.8). The biomass carbon density in the central region decreased significantly between 2005–2010 and 2015–2020, and the distribution of changes was more concentrated in 2015–2020. In contrast, there was a significant increase in distribution in the central region between 2010 and 2015, accompanied by a decrease in the eastern region.

2.3.4 Uncertainty Analysis

According to the spatial distribution maps of AGB and BGB uncertainty (Fig. 2.9), it was found that the spatial distribution pattern was similar to that of elevation, and it was also related to the geo-environmental factors such as climate, soil, and the number and quality of the measured sample points, and the highest uncertainty was found in the forested area in the southern part of the study area, and the prediction results had a high level of uncertainty due to the limitation of accessibility to forests, and the abundance of feature resources with a wide variety of species. For the spatial distribution of AGB uncertainty, the uncertainty was highest in the southern part of the study area, followed by the northern part of the study area due to the limitation of the measured sample points, and the uncertainty of prediction results was lower in the central part of the study area due to the concentration of sample points and the similar distribution of biomass values. For the spatial distribution of BGB uncertainty, the underground biomass of plateau permafrost in the western part of the study area was extremely low, and the uncertainty of the prediction results was low, followed by the central part, and the southern part had the highest uncertainty of the prediction results.

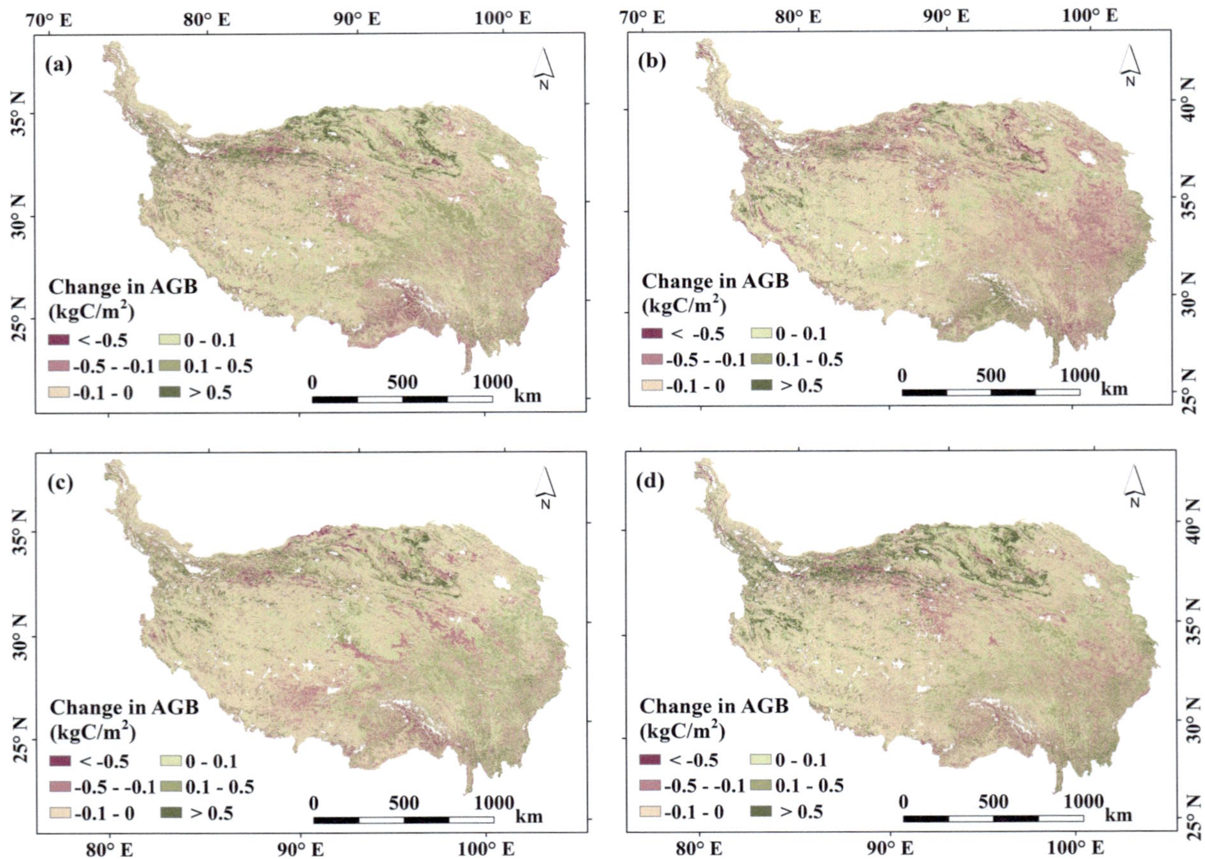

Fig. 2.7 Changes of aboveground biomass carbon density, **a** 2005–2010, **b** 2010–2015, **c** 2015–2020, **d** 2005–2020

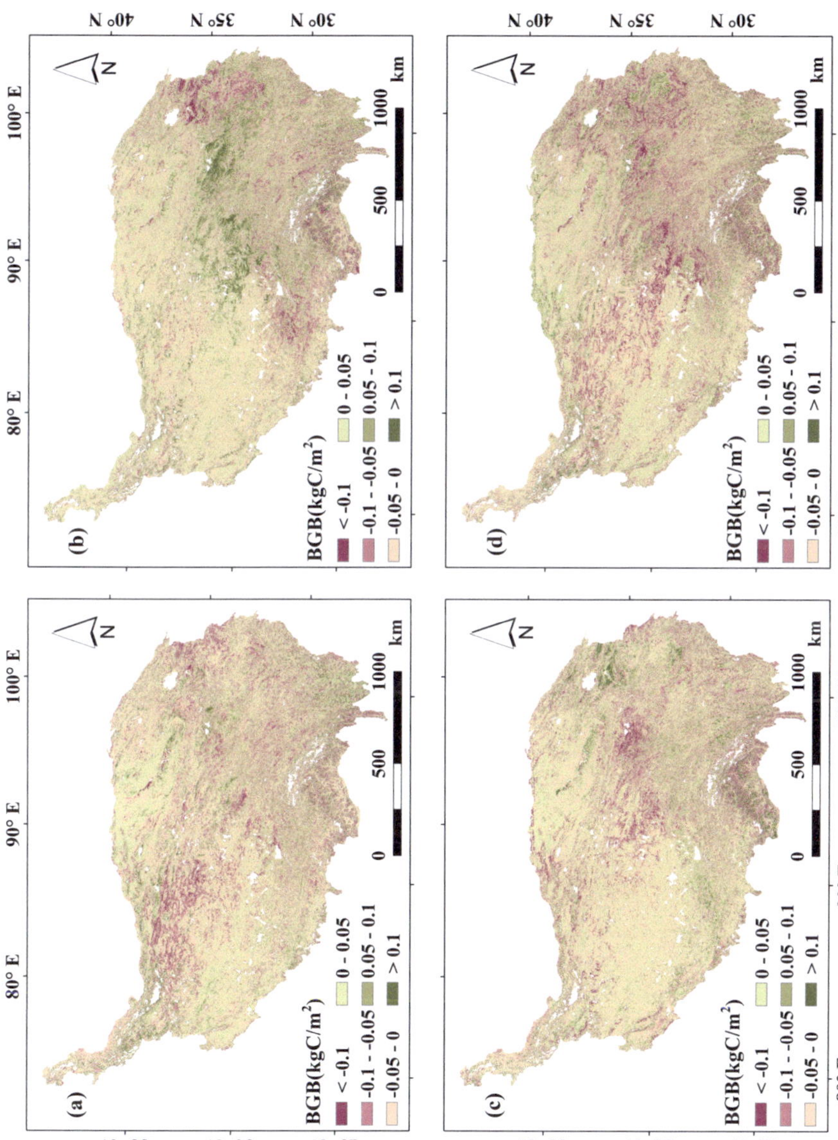

Fig. 2.8 Changes of belowground biomass carbon density, **a** 2005–2010, **b** 2010–2015, **c** 2015–2020, **d** 2005–2020

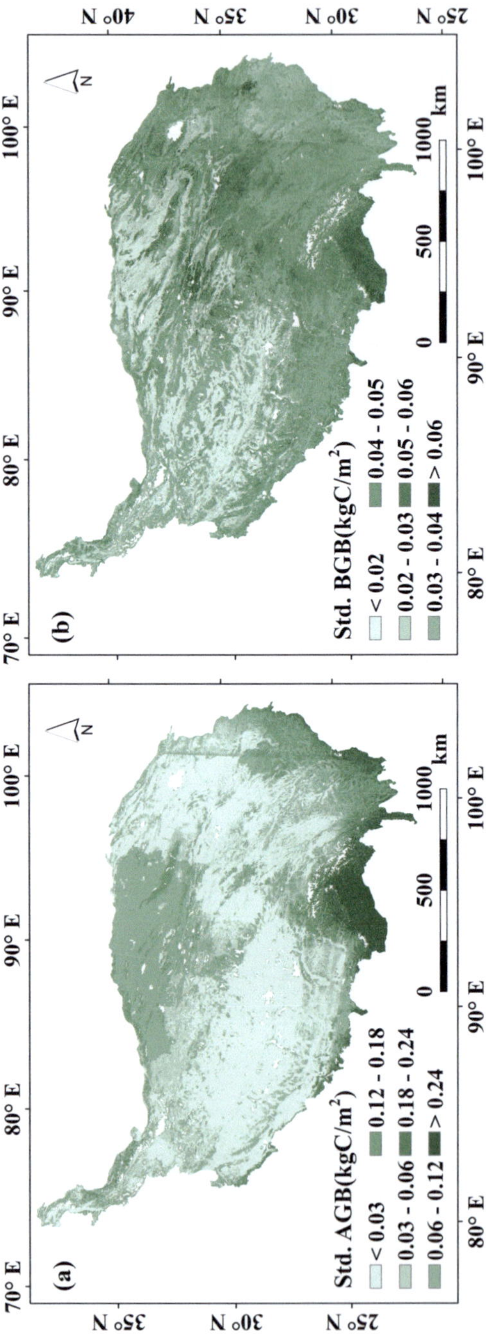

Fig. 2.9 Uncertainty analysis, **a** AGB, **b** BGB

2.4 Discussion

2.4.1 *Uncertainty of Biomass Remote Sensing Inversion*

The accuracy and uncertainty of biomass are mainly affected by several factors, such as sample point data, remote sensing images, environmental variable data and model selection, resulting in overestimation or underestimation of biomass (Lister et al., 2020; Zhou et al., 2021). First, the number of sampling points and their spatial distribution: there were 945 AGB and 764 BGB in the study area. In the field survey, the collection of ground samples was not only costly and labor-intensive, but also the sample points tended to be set up in the areas with single features and high accessibility, which resulted in the concentration of sampling points in the eastern part of the study area (Gao et al., 2020). In addition, the western part of the study area, which is higher and less accessible, has fewer sample points relative to the east-central region (Fig. 2.1), resulting in geographic bias in data coverage and exacerbating the uncertainty of model predictions. These limitations suggest the need to optimize the distribution of sample points and increase the number of sample points in future studies to improve the objectivity and accuracy of AGB estimation. Second, without considering the errors generated during the imaging process (environmental factors, sensor degradation, and erroneous image processing techniques), the remote sensing data errors are mainly manifested in the spatial and temporal consistency between the sampling points and the remote sensing data (Tian et al., 2002). Spatially, there is a difference between the low spatial resolution of the remote sensing data (1 km $\times$ 1 km) and the sampling points (1 m $\times$ 1 m or 5 m $\times$ 5 m), and this difference will produce a certain amount of error in the model (Eisfelder et al., 2012; Yuan et al., 2016). Temporally, sample point collection and laboratory processing are difficult and time-consuming, but a sufficient number of sample point data is needed for modeling as well as for validating the accuracy of the model. Therefore, it is difficult to have enough sample point data to match remote sensing data at the same time. Finally, comparing the estimation accuracy of biomass by different machine-learning models, there is no statistical difference in estimation accuracy among several commonly used machine-learning models (Han et al., 2022), and the model that performs optimally is different for different research subjects as well as data sources (Morais et al., 2021). Besides, the model accuracy is usually influenced by various factors, such as the geographic features of the study area, the distribution of sample points, and the selection of model parameters.

2.4.2 *Variable Importance Analysis*

Importance statistics of different environmental variables showed that soil factor accounted for 32% of the importance of aboveground biomass, followed by topographic factor (31%) and climatic factor (22%). On forest AGB estimation, Huang et al.'s study showed that the importance values of topographic and climatic factors were higher, but the importance value of soil factor was lower (Huang et al., 2023). Bennett et al.'s study also confirmed that climatic factor was more important than soil factor in the estimation of forest biomass (Bennett et al., 2020). But for different vegetation cover types, such as grassland and forest, the soil factor is more sensitive to the biomass response. The biological factors (NDVI, EVI) calculated in this paper are of low importance, which is largely influenced by the number of selected variables. Consideration can be given to adding spectral band variables as supplements and calculating more vegetation indices as input variables to the model (Li et al., 2020). Zhang et al.'s study considered 15 vegetation indices in the inversion of aboveground biomass of grasslands in Xinjiang, and the vegetation indices contributed significantly to the model in all types of grasslands (Zhang et al., 2023). In addition, the Tibetan Plateau is rich in grasslands, and human activities (grazing, etc.) have a large impact on AGB (Wang et al., 2023), which is also taken into account in paper (8% importance). Finally, texture feature is a visual feature that reflects the homogeneous phenomenon in the image, and Behera et al. have already used it as an influence factor to input into the model (Behera et al., 2024), while having the importance of 83.34% in the model. In future, textural features can be considered as environmental variables to estimate biomass.

2.4.3 Comparison with Other Products

In this study we used the RF model to estimate biomass and its spatial distribution, and compared the estimation results with AGB and BGB products produced by Han et al. (2022) (Fig. 2.10). It shows that its value is concentrated in the low value area. For the readability and aesthetics of the graph, we defined the range of aboveground biomass carbon density as 0–0.3 kg C/m^2. For aboveground biomass carbon density, there is some underestimation between the product and estimation values. The estimation results and product results have different degrees of underestimation compared with the measured point data. Although the product values have a large underestimation, they have a more accurate simulation effect in the low-value areas. For the carbon density of belowground biomass, the product values were consistent with the simulated values, but both of them underestimated the high value areas. In addition, the product values are more concentrated in low-value areas (Han et al., 2022).

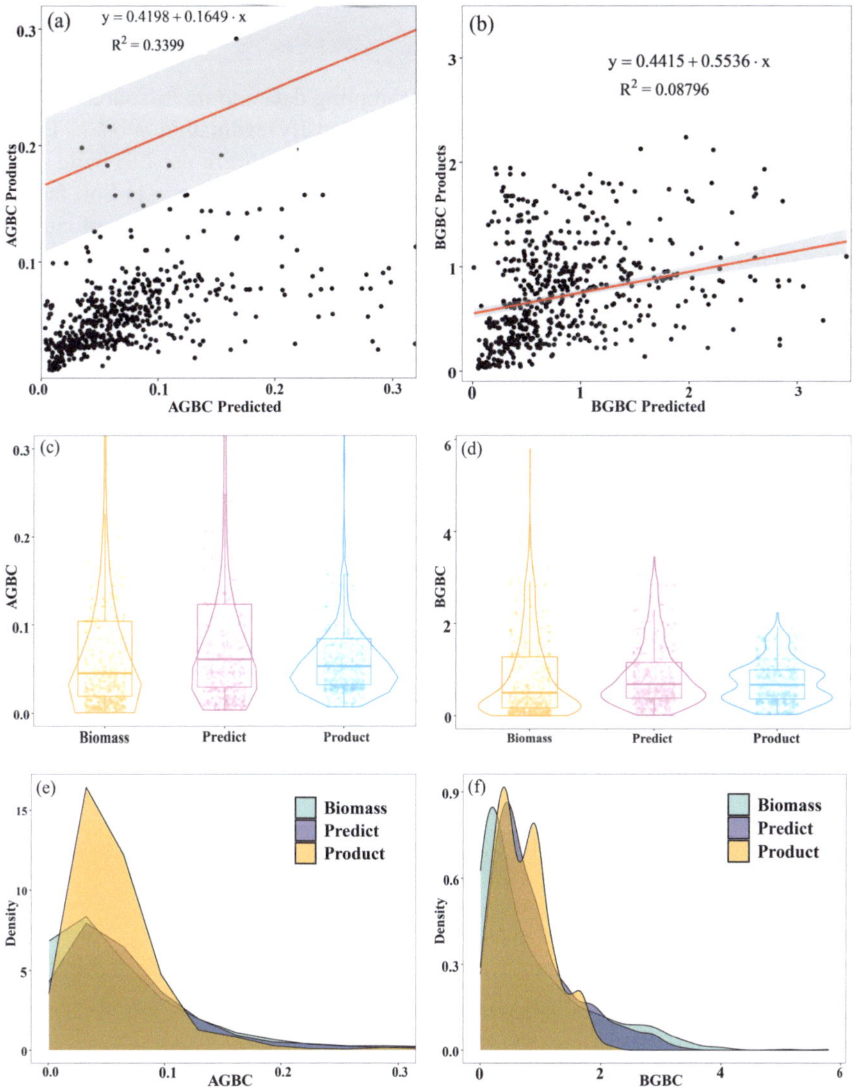

Fig. 2.10 **a** Comparison between predicted AGB carbon density and product, **b** comparison between predicted BGB carbon density in this study and product values, **c** comparison of field measured, predicted and product value of AGB carbon density, **d** comparison of field measured, predicted and product value of BGB carbon density, **e** density histograms of field measured, predicted and product values of AGB carbon density, **f** density histograms of field measured, predicted and product values of BGB carbon density

2.5 Conclusions

In this study, 945 AGB and 764 BGB field sampling data and multi-source remote variables were used to build biomass carbon density estimation model. Three machine-learning algorithms of RF, SVM, and XGBoost were used to build estimation model of biomass carbon density. The results showed that: (1) Soil factor, topographic factor, and climatic factor are the main influencing factors of carbon density changes in AGB and BGB. (2) For the spatial distribution estimation of biomass carbon density in the Tibetan Plateau, RF model obtained the highest accuracy for AGB ($R^2 = 0.519$, RMSE = 1.283) and BGB ($R^2 = 0.378$, RMSE = 0.705). (3) During 2001 and 2020, AGB carbon density showed a growth trend annually, and BGB carbon density varied fluctuating with the year, with no obvious growth or decrease trend. (4) The simulation results of AGB and BGB on the Tibetan Plateau were consistent in spatial distribution, with the overall characteristics of high in the east and low in the west.

References

Behera, D., Kumar, V. A., Rao, J. P., Padal, S. B., Ayyappan, N., & Reddy, C. S. (2024). Estimating aboveground biomass of a regional forest landscape by integrating textural and spectral variables of Sentinel-2 along with ancillary data. *Journal of the Indian Society of Remote Sensing, 52*, 917–929.

Bennett, A. C., Penman, T. D., Arndt, S. K., Roxburgh, S. H., & Bennett, L. T. (2020). Climate more important than soils for predicting forest biomass at the continental scale. *Ecography, 43*, 1692–1705.

Biau, G. (2012). Analysis of a random forests model. *The Journal of Machine Learning Research, 13*, 1063–1095.

Breiman, L. (2001). Random forests. *Machine Learning, 45*, 5–32.

Chen, T., & Guestrin, C. (2016). XGBoost: A scalable tree boosting system. In *Proceedings of the 22nd ACM SIGKDD International Conference on Knowledge Discovery and Data Mining* (pp. 785–794).

Eisfelder, C., Kuenzer, C., & Dech, S. (2012). Derivation of biomass information for semi-arid areas using remote-sensing data. *International Journal of Remote Sensing, 33*, 2937–2984.

Fawagreh, K., Gaber, M. M., & Elyan, E. (2014). Random forests: From early developments to recent advancements. *Systems Science and Control Engineering, 2*, 602–609.

Gao, Y., Skutsch, M., Paneque-Gálvez, J., & Ghilardi, A. (2020). Remote sensing of forest degradation: A review. *Environmental Research Letters, 15*, 103001.

Han, D., Hu, Z., Wang, X., Wang, T., Chen, A., Weng, Q., Liang, M., Zeng, X., Cao, R., & Di, K. (2022). Shift in controlling factors of carbon stocks across biomes on the Qinghai-Tibetan Plateau. *Environmental Research Letters, 17*, 074016.

Hari, M., & Tyagi, B. (2022). Terrestrial carbon cycle: Tipping edge of climate change between the atmosphere and biosphere ecosystems. *Environmental Science: Atmospheres, 2*, 867–890.

Hearst, M. A., Dumais, S. T., Osuna, E., Platt, J., & Scholkopf, B. (1998). Support vector machines. *IEEE Intelligent Systems and Their Applications, 13*, 18–28.

Huang, H., Liu, C., Wang, X., Zhou, X., & Gong, P. (2019). Integration of multi-resource remotely sensed data and allometric models for forest aboveground biomass estimation in China. *Remote Sensing of Environment, 221*, 225–234.

Huang, T., Ou, G., Wu, Y., Zhang, X., Liu, Z., Xu, H., Xu, X., Wang, Z., & Xu, C. (2023). Estimating the aboveground biomass of various forest types with high heterogeneity at the provincial scale based on multi-source data. *Remote Sensing, 15*, 3550.

Hurtt, G. C., Fisk, J., Thomas, R. Q., Dubayah, R., Moorcroft, P. R., & Shugart, H. H. (2010). Linking models and data on vegetation structure. *Journal of Geophysical Research: Biogeosciences, 115*, 2009GJ000937.

Jansen, V. S., Kolden, C. A., Taylor, R. V., & Newingham, B. A. (2016). Quantifying livestock effects on bunchgrass vegetation with Landsat ETM+ data across a single growing season. *International Journal of Remote Sensing, 37*, 150–175.

Keenan, T. F., & Williams, C. A. (2018). The terrestrial carbon sink. *Annual Review of Environment and Resources, 43*, 219–243.

Koju, U., Zhang, J., & Gilani, H. (2017). Exploring multi-scale forest above ground biomass estimation with optical remote sensing imageries. *IOP Conference Series: Earth and Environmental Science, 57*, 012011.

Kumar, L., & Mutanga, O. (2017). Remote sensing of above-ground biomass. *Remote Sensing, 9*, 935.

Kumar, L., Sinha, P., Taylor, S., & Alqurashi, A. F. (2015). Review of the use of remote sensing for biomass estimation to support renewable energy generation. *Journal of Applied Remote Sensing, 9*, 097696.

Li, Y., Li, M., Li, C., & Liu, Z. (2020). Forest aboveground biomass estimation using Landsat 8 and Sentinel-1A data with machine learning algorithms. *Scientific Reports, 10*, 9952.

Lister, A. J., Andersen, H., Frescino, T., Gatziolis, D., Healey, S., Heath, L. S., Liknes, G. C., McRoberts, R., Moisen, G. G., Nelson, M., Riemann, R., Schleeweis, K., Schroeder, T. A., Westfall, J., & Wilson, B. T. (2020). Use of remote sensing data to improve the efficiency of national forest inventories: A case study from the United States National Forest Inventory. *Forests, 11*, 1364.

Lu, D. (2006). The potential and challenge of remote sensing-based biomass estimation. *International Journal of Remote Sensing, 27*, 1297–1328.

Lu, D., Chen, Q., Wang, G., Liu, L., Li, G., & Moran, E. (2016). A survey of remote sensing-based aboveground biomass estimation methods in forest ecosystems. *International Journal of Digital Earth, 9*, 63–105.

Ma, W., Wu, T., Wu, X., Yue, G., Li, R., Li, X., Zhu, X., Hu, G., Qiao, Y., Hao, J., & Ni, J. (2021). Warming could shift steppes to carbon sinks and meadows to carbon sources in permafrost regions: Evidence from the improved IBIS model. *Catena, 200*, 105168.

Meng, B., Liang, T., Yi, S., Yin, J., Cui, X., Ge, J., Hou, M., Lv, Y., & Sun, Y. (2020). Modeling alpine grassland above ground biomass based on remote sensing data and machine learning algorithm: A case study in east of the Tibetan Plateau, China. *IEEE Journal of Selected Topics in Applied Earth Observations and Remote Sensing, 13*, 2986–2995.

Morais, T. G., Teixeira, R. F. M., Figueiredo, M., & Domingos, T. (2021). The use of machine learning methods to estimate aboveground biomass of grasslands: A review. *Ecological Indicators, 130*, 108081.

Nichol, J. E., & Sarker, Md. L. R. (2011). Improved biomass estimation using the texture parameters of two high-resolution optical sensors. *IEEE Transactions on Geoscience and Remote Sensing, 49*, 930–948.

Pan, Y., Ren, C., Liang, Y., Zhang, Z., & Shi, Y. (2020). Inversion of surface vegetation water content based on GNSS-IR and MODIS data fusion. *Satellite Navigation, 1*, 21.

Qin, L., Meng, S., Zhou, G., Liu, Q., & Xu, Z. (2021). Uncertainties in above ground tree biomass estimation. *Journal of Forestry Research, 32*, 1989–2000.

Sinha, S., Jeganathan, C., Sharma, L. K., & Nathawat, M. S. (2015). A review of radar remote sensing for biomass estimation. *International Journal of Environmental Science and Technology, 12*, 1779–1792.

Soenen, S. A., Peddle, D. R., Hall, R. J., Coburn, C. A., & Hall, F. G. (2010). Estimating aboveground forest biomass from canopy reflectance model inversion in mountainous terrain. *Remote Sensing of Environment, 114*, 1325–1337.

Stockmann, U., Adams, M. A., Crawford, J. W., Field, D. J., Henakaarchchi, N., Jenkins, M., Minasny, B., McBratney, A. B., Courcelles, V. D. R. D., Singh, K., Wheeler, I., Abbott, L., Angers, D. A., Baldock, J., Bird, M., Brookes, P. C., Chenu, C., Jastrow, J. D., Lal, R., et al. (2013). The knowns, known unknowns and unknowns of sequestration of soil organic carbon. *Agriculture, Ecosystems & Environment, 164*, 80–99.

Tian, X., Yan, M., Van Der Tol, C., Li, Z., Su, Z., Chen, E., Li, X., Li, L., Wang, X., Pan, X., Gao, L., & Han, Z. (2017). Modeling forest above-ground biomass dynamics using multi-source data and incorporated models: A case study over the qilian mountains. *Agricultural and Forest Meteorology, 246*, 1–14.

Tian, Y., Shi, Y., & Liu, X. (2012). Recent davances on support vector machines research. *Technological and Economic Development of Economy, 18*, 5–33.

Tian, Y., Woodcock, C., Wang, Y., Privette, J., Shabanov, N., Zhou, L., Zhang, Y., Buermann, W., Dong, J., & Veikkanen, B. (2002). Multiscale analysis and validation of the MODIS LAI product. Uncertainty assessment. *Remote Sensing of Environment, 83*, 414–430.

Wang, S., Tuya, H., Zhang, S., Zhao, X., Liu, Z., Li, R., & Lin, X. (2023). Random forest method for analysis of remote sensing inversion of aboveground biomass and grazing intensity of grasslands in Inner Mongolia, China. *International Journal of Remote Sensing, 44*, 2867–2884.

Wani, O. A., Kumar, S. S., Hussain, N., Wani, A. I. A., Babu, S., Alam, P., Rashid, M., Popescu, S. M., & Mansoor, S. (2023). Multi-scale processes influencing global carbon storage and land-carbon-climate nexus: A critical review. *Pedosphere, 33*, 250–267.

Wei, D., Qi, Y., Ma, Y., Wang, X., Ma, W., Gao, T., Huang, L., Zhao, H., Zhang, J., & Wang, X. (2021). Plant uptake of CO_2 outpaces losses from permafrost and plant respiration on the Tibetan Plateau. *Proceedings of the National Academy of Sciences, 118*, e2015283118.

Wu, D., Piao, S., Zhu, D., Wang, X., Ciais, P., Bastos, A., Xu, X., & Xu, W. (2020). Accelerated terrestrial ecosystem carbon turnover and its drivers. *Global Change Biology, 26*, 5052–5062.

Xu, L., He, N., & Yu, G. (2019). A dataset of carbon density in Chinese terrestrial ecosystems (2010s). *China Scientific Data, 4*, 21.86101/csdata.2018.0026.zh.

Yuan, X., Li, L., Tian, X., Luo, G., & Chen, X. (2016). Estimation of above-ground biomass using MODIS satellite imagery of multiple land-cover types in China. *Remote Sensing Letters, 7*, 1141–1149.

Zhang, R. P., Zhou, J. H., Guo, J., Miao, Y. H., & Zhang, L. L. (2023). Inversion models of aboveground grassland biomass in Xinjiang based on multisource data. *Frontiers in Plant Science, 14*, 1152432.

Zhang, W., Zhao, L., Li, Y., Shi, J., Yan, M., & Ji, Y. (2022). Forest above-ground biomass inversion using optical and SAR images based on a multi-step feature optimized inversion model. *Remote Sensing, 14*, 1608.

Zhang, Y., Liang, S., & Yang, L. (2019). A review of regional and global gridded forest biomass datasets. *Remote Sensing, 11*, 2744.

Zhou, W., Li, H., Xie, L., Nie, X., Wang, Z., Du, Z., & Yue, T. (2021). Remote sensing inversion of grassland aboveground biomass based on high accuracy surface modeling. *Ecological Indicators, 121*, 107215.

Chapter 3
Enhancing Vegetation and Soil Carbon Estimation Accuracy with HASM Data Fusion

Abstract Accurate and effective accounting of grassland aboveground biomass (AGB) is essential for grassland carbon storage accounting and pastoral agriculture sustainability. In this study, we combined AGB field survey data and remote sensing data to build a suitable model to estimate the grassland AGB in the three-river source region (TRSR) of China. Three machine learning models were used to simulate the grassland AGB from 2001 to 2019, including support vector machine (SVM), random forest (RF), and high-accuracy surface modeling (HASM). The results show that (1) the HASM achieved better results than the RF and SVM models ($R^2 = 0.8459 > 0.72 > 0.5858$; RMSE = $29 < 41 < 56$), and the HASM results reproduce the spatial distribution characteristics of the biomass well. The subsequent spatiotemporal analysis of the AGB conducted in this study was based on the results of the HASM. (2) The highest AGB was located in the eastern and central-southern parts of the TRSR, and the lowest AGB was distributed in the western region. (3) The overall change in the AGB revealed that the percentage of the area that experienced a significant increase in AGB (21%) was larger than that of the area that experienced a significant decrease (13%), and the stable areas accounted for 66% of the total area. The grassland AGB increased by 1 g/m^2/yr during 2001–2019. (4) The factors driving the changes in the AGB were analyzed. Overall, the warm and wet climate conditions promoted grass growth in most regions of the TRSR. (5) However, the biomass decreased in some regions with warm and wet conditions. For example, overgrazing and increased populations of grazing led to significant biomass decreases in the towns of Ziketang and Heka in Xinghai County. In this study, the grassland AGB was simulated based on the HASM model, with a high accuracy and a spatial resolution of 500 m. The results of this study provide a scientific basis for grassland resource protection and the highly effective implementation of grassland restoration projects in China.

Keywords High accuracy surface modeling · Machine learning · Aboveground biomass carbon · Soil organic carbon · Tibetan plateau

© The Author(s), under exclusive license to Springer Nature Singapore Pte Ltd. 2026

W. Zhou et al., *Carbon Sink Estimation and Evaluation of Ecological Function Changes on the Tibetan Plateau*, https://doi.org/10.1007/978-981-95-8664-6_3

3.1 Introduction

Above-ground biomass (AGB) constitutes the organic mass formed through photosynthesis and is an integral component in accounting of the global carbon cycle (Chopping et al., 2006; Nie et al., 2018; Scurlock et al., 2002). It is also an important parameter to evaluate land ecosystem structure and function (Zheng et al., 2004). Since the 1960s, vegetation biomass research has been a hot pot in terrestrial ecosystems. Based on field investigations, the dynamic changes and spatial distribution of vegetation biomass in terrestrial ecosystems have been explored (Miehle et al., 2010).

Grasslands are the largest terrestrial ecosystem on the Earth's surface, accounting for approximately 40% of the land area (Scurlock & Hall, 1998). The productivity accounts for about 20% of the total productivity of terrestrial ecosystems worldwide (O'Mara, 2012). Grassland ecosystem plays a crucial role in maintaining material circulation The balance between the environment and greenhouse gases plays an important role, especially in the research of global carbon storage and sink, which has significant implications (O'Mara, 2012). China has 4 million km^2 of grasslands, accounting for 40% of the total land area of China (Ren et al., 2011). Therefore, the amount of grassland AGB can significantly affect the number of grazing livestock, and thus, it can affect the livelihoods of pastoralists (Long, 2007). Therefore, a more accurate and operational accounting model for AGB estimation is urgently required.

For the vegetation biomass estimation, there are two methods. One is field measurements, this method can provide very high-accuracy biomass data, but it's time consuming and laborious and difficultly obtain the large scale biomass information (Xu et al., 2008). Another method is remote sensing retrieval, which is recognized as the most operational and effective approach for biomass estimation at large scales (Claverie et al., 2012), especially with the data booming of multiple-source remote sensing with high spatial–temporal resolution.

Two primary approaches for remote sensing-based biomass estimation are statistical regression and machine learning models (Ali et al., 2016). Among the machine learning algorithms, the random forest (RF) algorithm has been effectively employed in vegetation monitoring. Its strength lies in its ability to assemble regressions from diverse datasets and to manage complex, non-linear interactions, thereby mitigating the impact of noise and high-dimensional feature spaces (Vincenzi et al., 2011). In recent years, the random forest (RF) algorithm has been widely applied to estimate wheat biomass (Wang et al., 2016) as well as to monitor grassland biomass using remote sensing data (Mutanga et al., 2012; Ramoelo et al., 2015; Wang et al., 2017; Xia et al., 2018). The support vector machine (SVM) algorithm has been also widely used in vegetation biomass estimation (Mutanga et al., 2012; Xiu et al., 2019). The artificial neural network (ANN) has also been used to simulate vegetation biomass (Xie et al., 2009; Yang et al., 2018), although it has achieved high-accuracy results in simulating vegetation biomass, a significant limitation is the difficulty of selecting the best parameters during training. According to Wang et al. (2016), the random forest (RF) algorithm yielded more accurate wheat biomass estimates than both SVM and ANN. Furthermore, RF demonstrated greater robustness than the ANN model.

Soil organic carbon (SOC) is the largest carbon pool in the global terrestrial ecosystem, and the carbon storage capacity of 1 m thick soil is 2000 Gt (Lal, 2004), accounting for 50–80% of total terrestrial carbon (IPCC, 2016). Fluctuations in SOC, however small, can rapidly influence CO_2 concentrations, with profound implications for global climate change and carbon cycling (Gholizadeh et al., 2018). The stability of soil organic carbon (SOC) is intricately linked to crop growth and food security, as it directly affects soil health and fertility, thereby underpinning agricultural productivity (Zhao et al., 2018a, 2018b). Therefore, acquiring precise spatiotemporal data on soil organic carbon is critical for the targeted implementation of terrestrial carbon sequestration strategies.

Accurate quantification of soil organic carbon (SOC) relies on three primary methodologies: field surveys, process-based models, and data-driven machine learning (ML) approaches. While field surveys provide high local accuracy, they are time-intensive, laborious, and difficult to scale spatially (Wadoux et al., 2020). Process-based models, though valuable for simulating ecosystem dynamics, often face limitations in parameterization and the representation of complex ecological processes. In contrast, data-driven ML methods, particularly those integrating multi-source remote sensing data, offer a scalable and efficient alternative for SOC estimation.

Studies on remote sensing-driven SOC estimation often employ multi-source data as input factors to build models based on linear regression or non-linear machine learning. This multi-source approach provides a richer characterization of covariates representing the soil-forming environment. Multispectral remote sensing is the most widely used data-driven model for soil organic carbon. For instance, vegetation indices and spectral bands have been successfully incorporated into global SOC products like SoilGrids (Hengl et al., 2017) and the Chinese Soil Information Grid Dataset (Liu et al., 2022). Meanwhile, microwave remote sensing, particularly synthetic aperture radar (SAR), has shown growing potential in estimating soil properties. SAR is effective at characterizing vegetation-soil interactions (Ceddia et al., 2017) and has been applied for SOC estimation in regions such as China's Heihe River Basin (Qian et al., 2025; Zhou et al., 2020).

Process-based models are a fundamental tool for understanding and predicting changes in the Earth's surface systems. Widely applied in SOC simulation, models such as DNDC and Century can have their key drivers obscured by structural uncertainties. Furthermore, parameter sensitivity varies significantly across different ecosystems, necessitating model structural refinement and parameter optimization to improve reliability. Consequently, as each technique has complementary merits, a major research focus in digital soil mapping is now on how to effectively synergize the high-precision of ground measurements with the powerful spatial data mining capabilities of data-driven methods to achieve more accurate SOC simulations.

Therefore, in this study, we used high-accuracy surface modeling (HASM) operate data fusion using field measurement and remote sensing retrieval, to improve the estimation accuracy and spatial variability depicting in grassland above biomass and

SOC estimation. HASM method is grounded in the fundamental theorem of surfaces and was initially developed to resolve errors and multi-scale issues in Earth surface representation (Yue et al., 2004, 2007, 2016). It has since been widely applied in simulating various surface features, such as soil properties, climatic variables, digital elevation models, CO_2 remote sensing reflectance, and vegetation productivity (Liu et al., 2017; Shi et al., 2017; Sun et al., 2013; Zhao et al., 2014, 2018a, 2018b). To our knowledge, few studies have used HASM to estimate the grassland biomass and SOC estimation in Tibetan Plateau.

In this study, we selected the three-river source region (TRSR) as the study area. Because it is an important ecological barrier due to its huge grassland area and it is one of the most important areas for livestock grazing. However, the grassland ecosystem in this region has undergone degradation due to the impacts of climate change and anthropogenic activities (Jiang & Zhang, 2016; Zhou et al., 2017). Therefore, the timely and accurate accounting of the grassland biomass and the development of a reliable simulation model are necessary to provide scientific guidance for the sustainable utilization of the grassland resources and for carbon sink accounting in the TRSR.

The major objectives of this study are (1) to simulate the grassland AGB and SOC using ML models and multi-source data; (2) to verify whether the simulation accuracy and the expression of spatial heterogeneity can be improved after data fusion using HASM; (3) to explore the spatial–temporal patterns of the grassland AGB in the TRSR; and (4) to explore the driving mechanism and main driving factors of biomass changes and SOC spatial distribution.

3.2 Dataset Source and Methodology

3.2.1 Study Area

The three-river source region, situated in the hinterland of the Qinghai-Tibet Plateau (Fig. 3.1), spans longitudes from 100° 53′ E to 102° 16′ E and latitudes from 34° 05′ N to 34° 55′ N. It serves as the headwater area for China's three major rivers: the Yellow River, the Yangtze River, and the Lancang River. Encompassing 14 counties, the region's economy is predominantly based on animal husbandry. With an average elevation exceeding 4000 m, the area experiences a typical continental plateau climate. With an annual average temperature of – 5.6 to 4.9 °C and annual precipitation of 262.2–772.8 mm (mostly from June to September), the region is predominantly grassland. Grasslands span 287,951 km^2, covering 80% of the area. The short growth season (May to August) leads to low forage yield. Of the seven grassland types, alpine meadow (76%) and alpine steppe (23%) are dominant (Yang et al., 2018) (Fig. 3.1). Nevertheless, climate warming and anthropogenic activities have caused severe ecosystem degradation and fragility in the region (Jiang & Zhang, 2016).

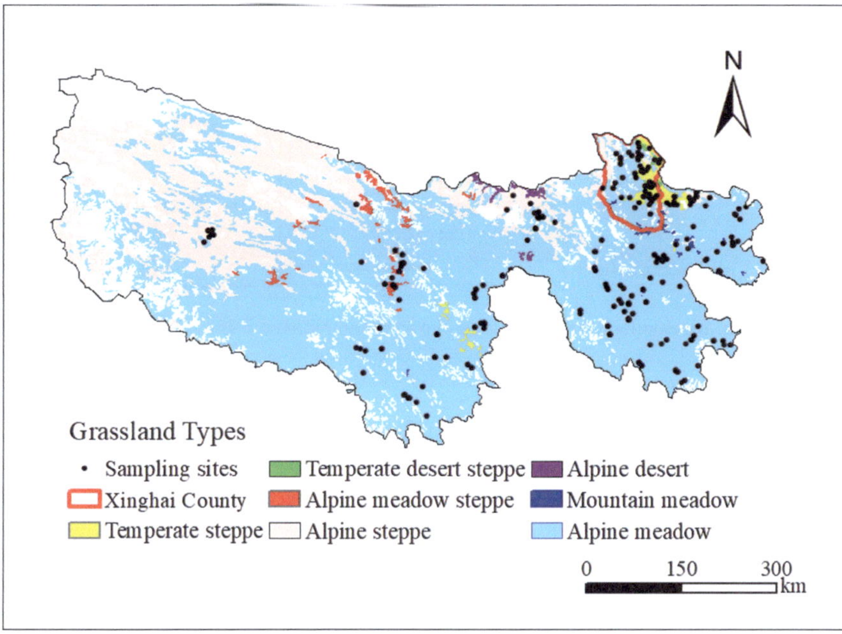

Fig. 3.1 Study area and field survey sites

3.2.2 Aboveground Biomass and Soil Sampling Data Source

The field survey data for grassland biomass was sourced from two initiatives: (1) 243 surveys were obtained from the Qinghai province monitoring center, gathered during the growing season (late July to August) from 2005 to 2015 (Fig. 3.1); and (2) 157 additional aboveground biomass (AGB) surveys were collected by our team from late July to early August in 2017, 2018, and 2019. The combined total of 400 AGB datasets was used for both model construction and accuracy assessment.

The field survey of AGB followed a standardized procedure. A rectangular site (3 km × 2 km) was established, with three plots distributed along its diagonal. Plot dimensions were set at 0.5 m × 0.5 m for homogeneous grassland and 1 m × 1 m for sparse or inhomogeneous vegetation. Comprehensive data—including spatial location, topographic factors, grassland attributes, fresh weight, and site conditions—were documented for each plot. Sample dry weight was measured after oven-drying at 65 °C for 48 h until a constant mass was achieved. To align with remote sensing data, the AGB values from plots falling within one MODIS pixel (500 m × 500 m) were averaged to represent the mean pixel AGB (Yang et al., 2018).

A total of 71 soil sampling sites were utilized to develop and validate the soil organic carbon (SOC) retrieval model (Fig. 3.2). Sampling was conducted during field surveys in July of 2019 and 2020. Based on Google Satellite imagery, sites were selected considering geographical features and accessibility. At each site, a composite

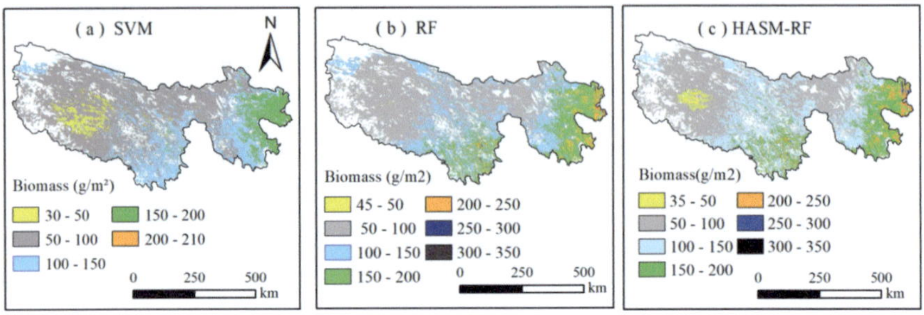

Fig. 3.2 Spatial distribution of the grassland aboveground biomass using the **a** SVM, **b** RF, and **c** HASM-RF

soil sample (0–20 cm depth) was obtained using a five-point mixing method. Detailed records—including geographic location, land use type, soil texture, vegetation type, and digital photographs—were documented for each site. Subsequently, the collected samples were air-dried, ground, and passed through a 0.25 mm sieve for laboratory analysis of SOC and soil texture.

3.2.3 Multiple Data and Preprocessing

3.2.3.1 Data for the AGB Estimation in 500 m Spatial Resolution

Grassland AGB was modeled based on the empirical relationship between field-surveyed AGB and satellite-derived NDVI. MODIS NDVI datasets from 2001 to 2019 at 500 m spatial resolution were employed. The data were processed as follows: the NDVI time series were smoothed using a Savitzky–Golay filter, and monthly maximum NDVI composites were generated from the 16-day interval data. All raster data were reprojected to the Albers Equal Area Conic system based on the WGS-84 datum.

A 90 m resolution digital elevation model (DEM) was acquired from the Shuttle Radar Topography Mission (SRTM, version 004) and resampled to 500 m to match the spatial consistency of other datasets. Using ArcGIS software, elevation, aspect, and slope were extracted from the reprojected (Albers Equal Area) DEM. Consequently, these topographic parameters—recognized for their impact on vegetation distribution and growth—were included in the biomass model, consistent with methodologies established in prior research (López et al., 2003; Xie et al., 2009).

Meteorological data, including average temperature, precipitation, and solar radiation (2001–2019), were sourced from the China Meteorological Science Data Center. These data were spatially interpolated into 500 m × 500 m grids using the Ordinary Kriging method. The resulting temperature and precipitation grids were subsequently utilized to develop the biomass model.

The model inputs in this study integrated three data categories: environmental factors, optical remote sensing, and SAR data. To test our first hypothesis, which posits that combining these diverse data types yields superior SOC estimation accuracy, we constructed and evaluated multiple models with varying data combinations.

The model was calibrated using nine environmental covariates sourced as follows: topographic indices (slope and aspect) were derived from a 12.5 m resolution ALOS DEM; climate variables (annual mean temperature and total precipitation) at 12.5 m resolution were provided by the Resources and Environmental Science and Data Center (RESDC); population density data was obtained from the Socioeconomic Data and Applications Center; and soil texture data was acquired from the Soil SubCenter of the National Earth System Science Data Center.

Five spectral bands from Sentinel-2A imagery and five spectral indices were used for modeling. The spectral indices included the Normalized Difference Vegetation Index (NDVI), Enhanced Vegetation Index (EVI), Clay Index (CI), Ratio Vegetation Index (RVI), and Soil-Adjusted Vegetation Index (SAVI), all of which were calculated based on reflectance in the visible, red, and near-infrared bands. The formulas for these vegetation indices, along with details on Sentinel-2 band data, with calculations referenced from previous studies SAR images were obtained from Sentinel-1A using Google Earth Engine (GEE) and were preprocessed. In this study, VV and VH polarization data were extracted from the descending pass on October 6, 2019, and the ascending pass on October 18, 2019. SAR data preprocessing includes radiometric calibration, speckle filtering, geometric correction, reprojection, and resampling, etc. All remote sensing data and environmental data were resampled to 90 m spatial resolution.

Five spectral bands from Sentinel-2A imagery and five spectral indices—NDVI, EVI, CI, RVI, and SAVI—were utilized for modeling. These indices were derived from reflectance in the visible, red, and near-infrared bands, with their calculation formulas referred to previous studies (Drusch et al., 2012). Additionally, Sentinel-1A SAR images were acquired and preprocessed via Google Earth Engine (GEE). VV and VH polarization data from both descending (October 6, 2019) and ascending (October 18, 2019) passes were extracted. Preprocessing steps for SAR data included radiometric calibration, speckle filtering, geometric correction, reprojection, and resampling. All remote sensing and environmental datasets were uniformly resampled to a 90 m spatial resolution.

3.2.3.2 Calculation of the Fractional Vegetation Cover

The vegetation cover (f_v) was defined as the projected area of the vegetation per unit of ground area. According to (Gutman & Ignatov, 1998), f_v is calculated based on the NDVI using a linear mixture model:

$$f_v = \frac{\mathrm{NDVI} - \mathrm{NDVI}_{\min}}{\mathrm{NDVI}_{\max} - \mathrm{NDVI}_{\min}} \tag{3.1}$$

where $NDVI_{min}$ is the minimum NDVI corresponding to bare land; and $NDVI_{max}$ is the maximum value for 100% coverage.

3.2.4 Methods

3.2.4.1 Machine Learning

SVM is evolved from robust statistical theory, which is well suited for handling small sample sizes, nonlinearity, and high-dimensionality problems. The support vector machine (SVM) model was implemented in R using the e1071 package, with a radial basis kernel applied through the caret interface. Model parameters were optimized via a grid search, resulting in a regularization parameter (C) of 1 and a gamma value of 0.1.

The random forest model, an ensemble algorithm originally developed by (Breiman, 2001), was implemented. Its ability to integrate multiple decision trees enhances predictive performance over single models (Hansen et al., 2006). Key parameters were determined through grid search, with the number of trees set at 600 and the maximum tree depth set at 7.

The extreme gradient boosting (XGBoost) model was employed, which integrates multiple decision trees through a boosting framework to enhance performance and mitigate overfitting. A grid search was conducted for parameter optimization, finalizing a learning rate of 0.4, maximum tree depth of 7, and 600 trees. Regularization was also applied to improve generalization. All models were developed and run in R.

3.2.4.2 HASM Model Description

HASM is a surface modeling method based on differential geometry, and it aims to solve the error and multi-scale problems that occur in surface modeling. A surface is uniquely defined by the first and the second fundamental coefficients. Thus, HASM was developed based on the fundamental theorem of surfaces (Yue et al., 2010). The main function of HASM satisfy the Gauss equation, by Taylor expansion of Gauss Codazzi equation. The above equation is described by Yue et al. (2007).

Combined with the information for the sampling points, the HASM can be transformed into the following least square problem:

$$\begin{cases} \min \left\| \begin{bmatrix} A \\ B \\ C \end{bmatrix} z^{(n+1)} - \begin{bmatrix} d^{(n)} \\ q^{(n)} \\ p^{(n)} \end{bmatrix} \right\|^2 \\ \text{s.t.} \\ S_1 z^{(n+1)} = k_1 \\ k_2 \le z^{(n+1)} \le k_3 \end{cases} \tag{3.2}$$

where, A, B, and C represent the coefficient matrices of the principal equation in the HASM method, where $z^{(n+1)}$ denotes the unknown quantity, and $d^{(n)}$, $q^{(n)}$, and $p^{(n)}$ are the right-hand term column vectors of the principal equation in the HASM method, with n being the iteration count. S_1 is the coefficient matrix of the sampling equation for optimizing control conditions in HASM, and k_1 is the right-hand term column vector of the sampling equation for optimizing control conditions. k_2 and k_3 are the upper and lower bound column vectors of $z^{(n+1)}$, respectively.

The HASM method was implemented using three input types: a driving field, optimal control conditions, and geographic information. The driving field was defined as the initial AGB simulations derived from either the RF or SVM models, while field-based AGB measurements served as the high-accuracy control conditions.

3.2.4.3 Estimation Accuracy Verification

To ensure robust model assessment, five-fold cross-validation was employed following a comprehensive comparison of methods in this study. Model performance was evaluated using five distinct metrics: the coefficient of determination (R^2), root mean square error (RMSE), mean absolute error (MAE), residual prediction deviation (RPD), and the ratio of performance to interquartile range (RPIQ). The mathematical definitions of these indices are provided in Eqs. (3.3) through (3.7):

$$R^2 = \left(\frac{\sum_{i=1}^{n}(O_i - \overline{O})(P_i - \overline{P})}{\sqrt{\sum_{i=1}^{n}(O_i - \overline{O})^2}\sqrt{\sum_{i=1}^{n}(P_i - \overline{P})^2}} \right)^2 \tag{3.3}$$

$$\text{RMSE} = \sqrt{\frac{1}{n}\sum_{i=1}^{n}(P_i - O_i)^2} \tag{3.4}$$

$$\text{MAE} = \frac{1}{n}\sum_{i=1}^{n}|P_i - O_i| \tag{3.5}$$

$$\text{RPD} = \frac{\text{SD}_\text{O}}{\text{RMSE}} \tag{3.6}$$

where, P_i and Oi represent the predicted and observed SOC content (g/kg) at location i, and n is the total number of sampling sites. $\overline{P}$ and $\overline{O}$ denote the mean value of predicted and observed SOC, respectively. Q1 and Q3 correspond to the first and third quartiles of the observed SOC, while SDO represents the standard deviation of the observed SOC.

3.2.4.4 Grassland Biomass Trend Analysis

Simple linear regression was used to analyze the change in the AGB for each pixel from 2001 to 2019. The calculation equations is as follows:

$$\theta_{\text{Slope}} = \frac{n \times \sum_{i=1}^{n} i \times C_i - \left(\sum_{i=1}^{n} i\right)\left(\sum_{i=1}^{n} C_i\right)}{n \times \sum_{i=1}^{n} i^2 - \left(\sum_{i=1}^{n} i\right)^2} \tag{3.7}$$

where n is the length of the study period (19 years); C_i is the aboveground biomass in year i; and θ_{Slope} is the slope of the trend line. To check the significance of the change in the biomass, a significance test (F-test) was conducted.

$$F = U \times \frac{n-2}{Q} \tag{3.8}$$

where U is the regression sum of squares; Q is the residual sum of squares; y_i is the simulated value of the biomass for year i and $\hat{y}_i$ is the regression value; $\overline{y}$ is the average value of the biomass from 2001 to 2019; and n is 19 years. According to the F-test, the change was classified into 6 categories: extremely significant increase ($P < 0.01$); extremely significant decrease ($P < 0.01$); significant increase ($P < 0.05$); significant decrease ($P < 0.05$); non-significant decrease ($P > 0.05$); and non-significant increase ($P > 0.05$).

3.3 Results

3.3.1 Accuracy Evaluation of Biomass Retrieval

Table 3.1 illustrates the comparison between field-surveyed aboveground biomass (AGB) and model-estimated values. The random forest (RF) model demonstrated superior performance to the support vector machine (SVM), with higher accuracy ($R^2 = 0.72$ vs. 0.59) and lower error (RMSE = 41 vs. 56). Based on these results, the spatial outputs from RF were selected as the initial field for the subsequent high accuracy surface modeling (HASM) process. The integrated HASM-RF model further

Table 3.1 Estimation accuracy comparison of AGB using different model

Model	R^2	RMSE	MAE	RPD
SVM	0.59	56	36	2.3
RF	0.72	41	32	3.1
RF-HASM	0.85	29	23	4.4

improved prediction performance ($R^2 = 0.85$, RMSE = 29), producing biomass estimates that more closely matched observed measurements compared to the RF-only approach.

3.3.2 Spatial Pattern of the Grassland AGB

Figure 3.2 illustrates the spatial distribution of mean annual grassland aboveground biomass (AGB) from 2001 to 2019, as simulated by the three models. The SVM model (Fig. 3.2a) generally underestimated AGB, particularly in the eastern part of the three-river source region (TRSR), where the maximum value did not exceed 210 g/m^2, while the western part exhibited extensive areas with the lowest biomass values. These spatial patterns are consistent with the quantitative accuracy assessment presented in Table 3.1. In contrast, the RF model (Fig. 3.2b) led to a systematic overestimation of AGB, with the western TRSR showing only limited areas of minimal biomass. The HASM-RF model (Fig. 3.2c) effectively compensated for the limitations of both the SVM and RF models, yielding simulations that better represent the actual magnitude and spatial variability of grassland AGB. As a result, all subsequent spatiotemporal analyses were conducted based on the HASM-RF model outputs.

Based on the HASM-RF simulations, the average grassland aboveground biomass (AGB) in the three-river source region (TRSR) from 2001 to 2019 was 117 g/m^2. Spatially, higher AGB values were predominantly distributed in the eastern and southern parts of the region, with a maximum value approaching 350 g/m^2 observed in the east. In contrast, the western part exhibited lower biomass levels, particularly near Ge'ermu City, where the minimum value dropped to 39 g/m^2. Moderately high AGB values of approximately 150 g/m^2 were found in central TRSR—including Dari, Qumalai, Zaduo, and Chengduo counties—as well as in the easternmost part of Zhiduo County. Overall, AGB values ranging from 50 to 100 g/m^2 were the most widespread across the study area.

3.3.3 *Annual Change Trend Analysis of AGB Based on HASM-RF from 2001 to 2019*

The grassland AGB increased by 1 $g/m^2/yr$ from 2001 to 2019. The statistical analysis revealed that the percentage of grassland area that experienced a significant increase in AGB (21.6%) was greater than the percentage that experienced a significant decrease (12.2%), and the stable area accounted for 66.1% of the total grassland area.

In terms of the spatial distribution, the AGB experienced the most significant increase in the far eastern regions of the study area (i.e., Zeku, Henan, Maqin, and Gande counties). Moreover, the regions with extremely significant increases were located in areas with high AGB values (Fig. 3.3), which indicates that the effects of grassland restoration was significant in these areas. The AGB experienced the most significant decrease in the northwestern parts of Qu Malai and Zhiduo counties (Fig. 3.3), indicating that the grassland degradation in these areas was significant. The far western regions experienced non-significant decreases; and the southern region of the TRSR experienced a non-significant increase.

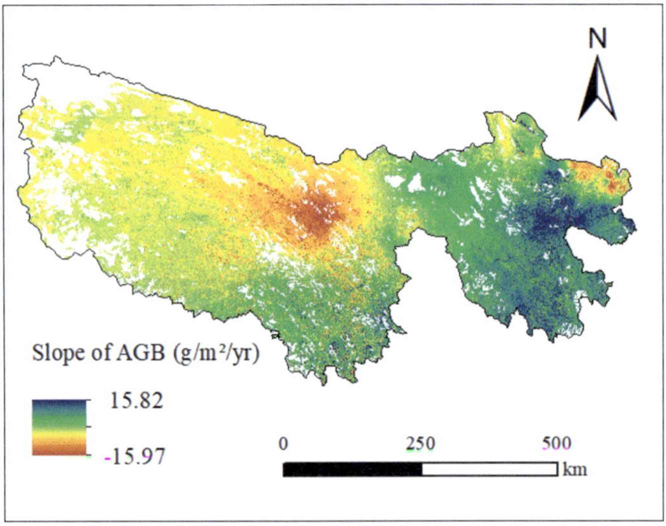

Fig. 3.3 Changes in the AGB from 2001 to 2019 in the TRSR

3.3.4 *Model Accuracy Comparison for SOC Estimation Based on Different Data Combination*

Correlation analysis were conducted to examine the relationships between SOC and various environmental covariates, optical data, and SAR data. For the individual data type, environmental covariates obtained higher estimation accuracy than using optical variables and SAR polarization data. Besides, estimation accuracy with optical variables achieved higher prediction accuracy than SAR Data. Among the three ML, the RF model obtained the highest R^2 value (0.58) and the lowest MAE (5.36) (Table 3.2).

Results indicated that model predictions were most accurate when multiple data sources were integrated. Both the combination of environmental with optical data and

Table 3.2 Comparison of SOC content estimation accuracy using different data combination

Data combination	Model	Calibration			Verification			
		R^2	RMSE	MAE	R^2	RMSE	MAE	RPD
Environmental	SVM	0.71	8.47	5.75	0.57	8.78	7.18	1.6
	RF	0.92	5.18	4.06	0.62	8.43	7.12	1.66
	XGBoost	0.95	3.45	2.66	0.69	8.11	6.72	1.78
Optical data	SVM	0.68	6.1	4.5	0.58	8.58	6.35	1.71
	RF	0.87	6.13	4.48	0.58	6.42	5.36	1.5
	XGBoost	0.92	4.67	3.22	0.47	7.41	6.31	1.42
SAR	SVM	0.65	8.29	6.64	0.2	8.77	7.43	1.02
	RF	0.83	7.35	5.8	0.3	8.69	7.82	1.07
	XGBoost	0.99	0.99	0.29	0.21	9.84	8.76	1.12
Environmental + Optical data	SVM	0.73	5.15	4.15	0.6	5.25	4.42	1.73
	RF	0.92	5.04	3.67	0.66	5.66	4.82	1.77
	XGBoost	0.97	2.58	1.84	0.66	6.16	5.24	1.6
Optical data + SAR	SVM	0.71	6.08	4.67	0.5	6.21	5.25	1.49
	RF	0.88	5.84	4.23	0.61	6.14	5.11	1.65
	XGBoost	0.99	0.99	0.27	0.63	6.72	5.58	1.49
Environmental + SAR	SVM	0.81	7.48	4.83	0.6	8.18	6.91	1.62
	RF	0.92	5.24	4.16	0.54	8.22	7.33	1.48
	XGBoost	0.99	0.99	0.28	0.69	6.42	5.52	1.71
Environmental + Optical + SAR data	SVM	0.79	6.21	4.18	0.56	6.71	5.71	1.48
	RF	0.92	4.96	3.61	0.59	6.68	5.48	1.68
	XGBoost	0.99	0.99	0.28	0.72	6.59	5.51	1.66

Data I: Environmental covariates; Data II: Optical images; Data III: SAR; Data IV: Environmental covariates and optical images; Data V: Optical images and SAR; Data VI: Environmental covariates and SAR; Data VII: Environmental covariates, optical images and SAR environmental with SAR data surpassed the predictive performance of any individual dataset. For example, using the environmental-SAR combination in RF and XGBoost models notably improved R^2 values to 0.54 and 0.69, far exceeding the results from SAR data alone (0.30 and 0.21). The highest accuracy was obtained by combining all three data types, leading to R^2 increases of 3% to 229% and RMSE reductions of 11% to 33% across the models (Table 3.2). Therefore, SAR data not only serve as a valuable input but can achieve performance on par with optical data in SOC estimation.

Overall, among the three machine learning models, XGBoost consistently achieved the highest estimation accuracy across all data combinations, followed by RF and SVM. However, there was a general underestimation of high SOC values across all models.

3.3.5 Spatial Distribution Analysis of SOC

The spatial distribution of soil organic carbon (SOC) in Xinghai County was mapped using dataset VII and three machine learning models: support vector machine (SVM), random forest (RF), and XGBoost (Fig. 3.4). The mean SOC content estimated by these models was 31.08 g/kg, 43.19 g/kg, and 27.57 g/kg for SVM, RF, and XGBoost, respectively. Compared to field-measured values, both the SVM and RF models overestimated SOC levels, while the XGBoost model produced slightly lower estimates. Overall, the SVM and RF models tended to underestimate areas with high SOC content and overestimate those with low SOC content (Fig. 3.4).

Spatially, the highest estimated SOC values were concentrated in the southernmost part of the study area—a region characterized by the highest humidity and most extensive forest coverage. This spatial distribution pattern corresponds closely with the regional variations in soil types and vegetation cover. Statistical results indicated that grass-felt soil, predominantly found in alpine meadows with an aboveground

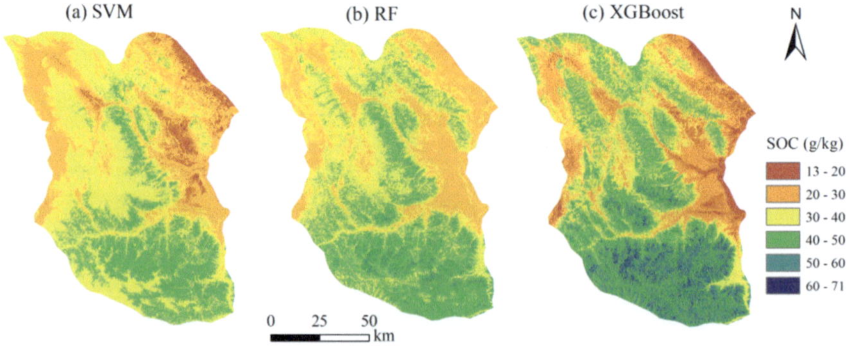

Fig. 3.4 Spatial distribution of predicted SOC using **a** SVM, **b** RF, and **c** XGBoost

Table 3.3 Comparison of SOC content prediction accuracy after data fusion using the HASM

Model combination	R^2	RMSE	MAE	RPD	RPIQ
SVM	0.56	6.71	5.71	1.48	1.76
RF	0.59	6.68	5.48	1.68	2.27
XGBoost	0.72	6.59	5.51	1.66	1.75
SVM + HASM	0.66	6.55	5.26	1.72	1.96
RF + HASM	0.72	5.87	5.24	1.95	2.68
XGBoost + HASM	0.85	6.26	5.15	1.84	2.58

biomass (AGB) of 130 g/m^2, showed the highest SOC content (50.8 g/kg). In contrast, chestnut soil, which mainly occurs in temperate grasslands and temperate desert grasslands (with AGB values of 120 g/m^2 and 84 g/m^2, respectively), exhibited the lowest SOC level (21.5 g/kg). Overall, the spatial heterogeneity of SOC across Xinghai County is largely consistent with the distribution of land cover and soil types.

3.3.6 *Simulation Accuracy Improvement After Data Fusion Using HASM*

The SOC spatial pattern after HASM data fusion was similar to the XGBoost results, but offered a more accurate representation of spatial details. It corrected overestimation in eastern and western Xinghai to reflect chestnut soil and alpine grassland, and adjusted underestimation in southern and central areas to align with grass-felt soil and alpine meadow. However, since these improvements were minor, it demonstrates that the XGBoost model alone already provides a reliable SOC simulation for this alpine environment for the alpine ecosystem. The data fusion process significantly improved the simulation accuracy. The R^2 values for SVM, RF, and XGBoost model increased by 18%, 22%, and 18%, respectively. Meanwhile, RMSE values declined by 2%, 12%, and 5%, respectively (Table 3.3).

3.4 Discussion

3.4.1 *Driving Mechanism Analysis of Grassland AGB Change*

A significant warming and wetting trend characterized the climate in the three-river source region (TRSR) from 2001 to 2019 (Fig. 3.5), consistent with the broader climatic shift observed across northwestern China since the 1980s (Shi, 2003). This

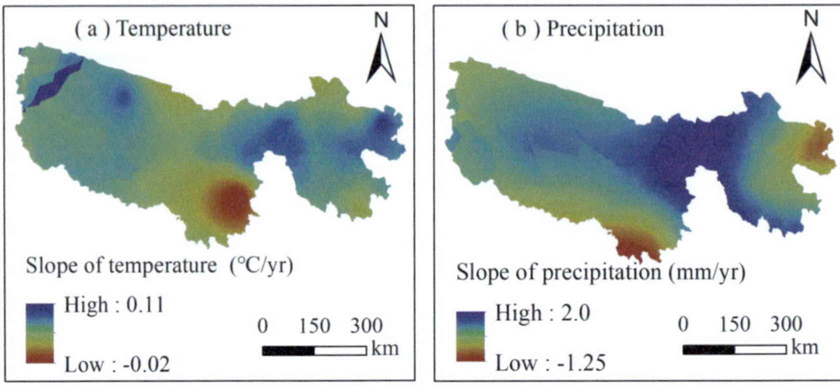

Fig. 3.5 The change trend of temperature (**a**) and precipitation (**b**) of TRSR during 2001–2019

generally favorable climate promoted an increase in aboveground biomass (AGB), in line with previous studies (Zheng et al., 2006). However, AGB declined in some areas despite the conducive warm-wet conditions. This anomaly can likely be attributed to the compounding effects of high elevation and low temperatures, which inherently limit plant growth; excessive rainfall may further intensify such cold stress (Mu et al., 2013). In addition, intensive human activities—such as overgrazing and population growth in parts of Xinghai County (e.g., Ziketang and Heka towns)—imposed significant pressure on local ecosystems, resulting in pronounced AGB losses in these regions.

3.4.2 The Impact of Human Activities on the AGB in Xinghai County

Xinghai County was selected as the study area for its diverse grassland types and comprehensive animal husbandry data. Aboveground biomass (AGB) increased across most of the county, despite notable decreases in Zi Ketan Town and scattered parts of Tang Naihai and Heka towns.

Since 2002, both the total and rural populations have risen, with particularly marked growth in Zi Ketan, Heka, and Tang Naihai towns. Concurrently, livestock numbers in Xinghai County expanded rapidly in 2003. This surge may reflect herders' attempts to secure a final opportunity to raise incomes through intensified grazing—a practice that preceded the implementation of China's Returning Grazing Land to Grassland Program (RGGP) later that same year. Under the RGGP, measures such as grazing bans, rotational grazing, and fence construction were introduced to alleviate pressure on degraded grasslands (Tong et al., 2004). Extensive research has since demonstrated the program's positive impacts. In a previous study, we found that

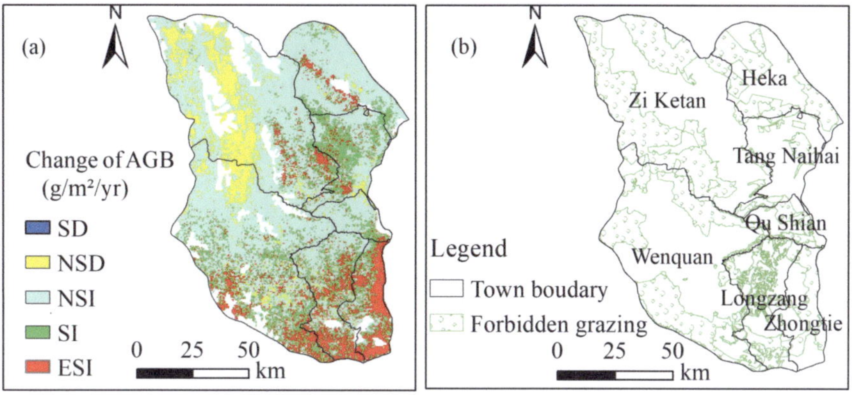

Fig. 3.6 Grassland AGB change in Xinghai during 2001–2019 (**a**), and the forbidden grazing in 2012 (**b**). *Note* SD is significant decrease, NSD is no significant decrease, NSI is no significant increase, SI is significant increase, ESI is extremely significant increase

human activities—primarily restoration policies—were the dominant driver of grassland recovery from 1982 to 2010, accounting for 94% of improvement in Qinghai (Zhou et al., 2017).

In Zhongtie and Wenquan towns, livestock numbers declined annually while the human population remained stable since 2002. The resultant reduction in human pressure, combined with a warmer and wetter climate, consequently promoted grassland growth and led to a significant increase in AGB.

The rural population grew markedly in Zi Ketan and Heka towns, while livestock numbers remained stable. Against this backdrop, the establishment of extensive forbidden grazing areas (994,073 m^2 and 999,172 m^2 in 2012, accounting for 50% and 55% of total grassland area, respectively (Fig. 3.6) sharply reduced the available pasture. This mismatch precipitated more intense overgrazing in the remaining areas, and the resulting aggravation of human acclivities pressures was a direct outcome of this policy-livestock imbalance, ultimately leading to reduced AGB.

3.4.3 Advantages for Data Fusion Using HASM and the Future Improvements

HASM improves grassland AGB and SOC estimation through two principal mechanisms. Its data fusion framework incorporates both point measurements and surface data, thereby producing results with greater precision than those derived from any single data source (Yue et al., 2016). Furthermore, the method significantly refines the depiction of spatial heterogeneity, resulting in more accurate representations of the variable's spatial distribution (Shi et al., 2017; Sun et al., 2013).

These advantages have been confirmed in this study, for grassland AGB and SOC estimation in Tibetan Plateau. For grassland AGB, the accuracy of the HASM-RF model ($R^2 = 0.8459$) was higher than that using only the RF model ($R^2 = 0.72$), and achieved in previous studies of the TRSR (Yan et al., 2015; Zeng et al., 2017). Furthermore, HASM more faithfully captures the spatial distribution of AGB compared to SVM and RF, especially at distribution extremes. This is evident in Fig. 3.4, which shows that SVM underestimates AGB in the low-biomass western TRSR, whereas RF yields overestimations in the east.

For SOC estimation, data fusion using HASM firstly obtained higher estimation accuracy, the R^2 values of the SVM, RF, and XGBoost models increased by 18%, 22%, and 18%, respectively, while their RMSE values decreased by 2%, 12%, and 5%. The spatial distribution patterns of SOC were also significantly improved, with reduced overestimation in high-SOC areas and underestimation in low-SOC areas compared to the results generated by any single model. Specifically, SVM and RF underestimated SOC values in central and southern Xinghai County but overestimated them in the eastern and western regions (Fig. 3.4). These improvements can be attributed to the ability of HASM to integrate both point and surface data, Moreover, HASM effectively captures the spatial heterogeneity of SOC distribution.

Although the AGB and SOC estimation using data fusion based on HASM model achieved a good performance in this study, it can still be further optimized by obtaining the following measurements. Firstly, the geographical factors, such as topography, grassland type, and soil type, can affect the grassland biomass and SOC. As shown in our previous study, SOC simulation accuracy based on different land cover types was superior to using any single model for the entire study area, with R^2 increasing by 11–15%, and RMSE decreasing by 7–45%. In future studies, the geographical factors must be considered. Secondly, the problem of the sparse sampling sites in the western part of the study area should be solved, and the accuracy can be improved accordingly. At present, the field sampling plots are distributed in the eastern and central regions of the TRSR. Thirdly, spatial–temporal inconsistencies between the sampling sites and the satellite data (Eisfelder et al., 2012), because large disparity inevitably produces modeling errors (Yuan et al., 2016).

3.4.4 Comparison with Other SOC Products

The most commonly used SOC products include the Global Harmonized World Soil Database (HWSD), SoilGrids with a 250 m spatial resolution for the World (Hengl et al., 2017), and SoilGrids with a 1 km resolution for China (Liu et al., 2022).

The spatial distribution of three SOC products were showed compared (Fig. 3.7). It revealed that the SoilGrids 250 m resolution product somewhat overestimated SOC content in areas with high SOC values, whereas the HWSD markedly underestimated SOC content across the entire region. In contrast, the SoilGrids 1 km product for China demonstrated higher SOC prediction accuracy than both SoilGrids 250 m and HWSD (Fig. 3.7). A key factor contributing to these discrepancies is likely the

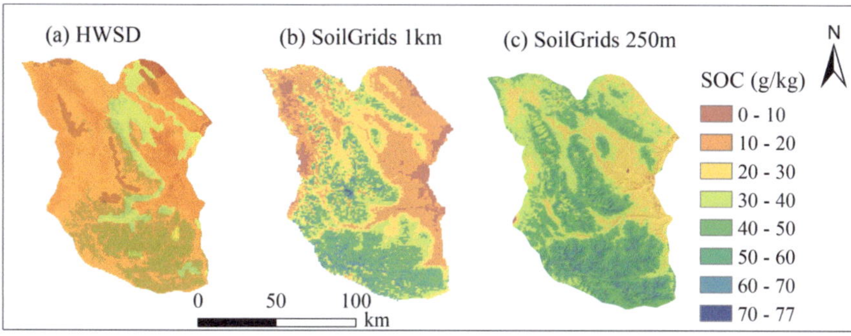

Fig. 3.7 Spatial distribution of SOC content for **a** HWSD, **b** SoilGrids at 1 km resolution, and **c** SoilGrids at 250 m resolution

timing of the underlying soil surveys. The SoilGrids 1 km product incorporated data collected between 2010 and 2018, which is temporally closer to the soil sampling conducted in this study from 2019 to 2020. In comparison, the HWSD for China relies on data from the 1980s, and the SoilGrids 250 m product uses data spanning the 1980s and 2000s (Hengl et al., 2017). The substantial time lag in these datasets likely led to differences in SOC estimates. Furthermore, the SoilGrids 1 km product benefited from environmental variables and remote sensing data that were derived from national-scale sources with finer spatial and thematic accuracy.

3.5 Conclusions

In this study, field survey biomass and SOC data, environmental covariates, multi-source remote sensing data, and three machine learning models were used to estimate AGB and SOC content. The results showed that: (1) The simulation accuracy of the RF (R^2 = 0.72) was greater than that of the SVM (R^2 = 0.5858). (2) Data fusion of RF-HASM obtained the best simulation accuracy (R^2 = 0.8459, RMSE = 29). (3) The mean grassland AGB during 2001–2019 was 117 g/m^2. The highest AGB values were located in the eastern and central-southern parts of the study area, and the lowest AGB values were located in the western part of the TRSR. Overall, the percentage of the study area that experienced a significant increase in AGB (21%). (4) The combination of environmental covariates, optical, and SAR remote sensing data resulted in the best estimation accuracy, with XGBoost outperforming both RF and SVM models; (3) after data fusion based on the HASM model, the estimation accuracy improved substantially, with R^2 increasing by 18–25%, and RMSE declining by 2–12%. Additionally, the spatial distribution of SOC content became more reasonable, with both overestimation and underestimation of AGB and SOC being corrected; and (4) compared to other SOC products, the estimation effectiveness of SOC is superior in this study. Therefore, this study supports our research hypothesis that data fusion based on the HASM model leads to better AGB and SOC estimation results. This research provides an accurate and novel method for SOC assessment, which can be widely applied in the production of national or global SOC data products and will offer a scientific foundation for the spatial planning of terrestrial carbon sequestration strategies.

References

Ali, I., Cawkwell, F., Dwyer, E., Barrett, B., & Green, S. (2016). Satellite remote sensing of grasslands: From observation to management–a review. *Journal of Plant Ecology, 9*, 1–23.

Breiman, L. (2001). Random Forests. *Machine Learning, 45*, 5–32.

Ceddia, M. B., Gomes, A. S., Vasques, G. M., & Pinheiro, E. F. M. (2017). Soil carbon stock and particle size fractions in the central amazon predicted from remotely sensed relief, multispectral and radar data. *Remote Sensing, 9*, rs9020124.

Chopping, M., Su, L., Laliberte, A., Rango, A., Peters, D. P. C., & Kollikkathara, N. (2006). Mapping shrub abundance in desert grasslands using geometric-optical modeling and multi-angle remote sensing with CHRIS/Proba. *Remote Sensing of Environment, 104*, 62–73.

Claverie, M., Demarez, V., Duchemin, B. T., Hagolle, O., Ducrot, D., Marais-Sicre, C., Dejoux, J. O., Huc, M., Keravec, P., & Béziat, P. (2012). Maize and sunflower biomass estimation in southwest France using high spatial and temporal resolution remote sensing data. *Remote Sensing of Environment, 124*, 844–857.

Drusch, M., Del Bello, U., Carlier, S., Colin, O., Fernandez, V., Gascon, F., Hoersch, B., Isola, C., Laberinti, P., Martimort, P., Meygret, A., Spoto, F., Sy, O., Marchese, F., & Bargellini, P. (2012). Sentinel-2: ESA's optical high-resolution mission for GMES operational services. *Remote Sensing of Environment, 120*, 25–36.

Eisfelder, C., Kuenzer, C., & Dech, S. (2012). Derivation of biomass information for semi-arid areas using remote-sensing data. *International Journal of Remote Sensing, 33*, 2937–2984.

Gholizadeh, A., Žižala, D., Saberioon, M., & Borůvka, L. (2018). Soil organic carbon and texture retrieving and mapping using proximal, airborne and Sentinel-2 spectral imaging. *Remote Sensing of Environment, 218*, 89–103.

Gutman, G., & Ignatov, A. (1998). The derivation of the green vegetation fraction from NOAA/AVHRR data for use in numerical weather prediction models. *International Journal of Remote Sensing, 19*, 1533–1543.

Hansen, J., Sato, M., Ruedy, R., Lo, K., Lea, D. W., & Medinaelizade, M. (2006). Global temperature change. *Proceedings of the National Academy of Sciences of the United States of America, 103*, 14288–14293.

Hengl, T., Mendes De Jesus, J., Heuvelink, G. B. M., Ruiperez Gonzalez, M., Kilibarda, M., Blagotić, A., Shangguan, W., Wright, M. N., Geng, X., Bauer-Marschallinger, B., Guevara, M. A., Vargas, R., MacMillan, R. A., Batjes, N. H., Leenaars, J. G. B., Ribeiro, E., Wheeler, I., Mantel, S., & Kempen, B. (2017). SoilGrids250m: Global gridded soil information based on machine learning. *PLoS ONE, 12*, e169748.

IPCC (2016). *Intergovernmental panel on climate change (IPCC)*. http://www.ipcc.ch/.

Jiang, C., & Zhang, L. (2016). Ecosystem change assessment in the Three-river Headwater Region, China: Patterns, causes, and implications. *Ecological Engineering, 93*, 24–36.

Lal, R. (2004). Soil carbon sequestration impacts on global climate change and food security. *Science, 304*, 1623–1627.

Liu, Y., Guo, J., Yue, T., & Zhao, N. (2017). *Simulation and analysis of carbon dioxide concentration in the surface layer*.

Liu, F., Wu, H. Y., Zhao, Y. G., Li, D. C., Yang, J. L., Song, X. D., Shi, Z., Zhu, A. X., & Zhang, G. L. (2022). Mapping high resolution National Soil Information Grids of China. *Science Bulletin, 67*, 328–340.

Long, R. J. (2007). Function of ecosystem in the Tibetan Grassland. *Science & Technology Review, 25*, 26–29.

López, I. F., Lambert, M. G., Mackay, A. D., & Valentine, I. (2003). The influence of topography and pasture management on soil characteristics and herbage accumulation in hill pasture in the North Island of New Zealand. *Plant and Soil, 255*, 421–434.

Miehle, P., Grote, R., Battaglia, M., Feikema, P. M., & Arndt, S. K. (2010). Evaluation of a process-based ecosystem model for long-term biomass and stand development of Eucalyptus globulus plantations. *European Journal of Forest Research, 129*, 377–391.

Mu, S. J., Yang, H. F., Li, J. L., Chen, Y. Z., Gang, C. C., Zhou, W., & Ju, W. M. (2013). Spatio-temporal dynamics of vegetation coverage and its relationship with climate factors in Inner Mongolia, China. *Journal of Geographical Sciences, 23*, 231–246.

Mutanga, O., Adam, E., & Cho, M. A. (2012). High density biomass estimation for wetland vegetation using WorldView-2 imagery and random forest regression algorithm. *International Journal of Applied Earth Observation and Geoinformation, 18*, 399–406.

Nie, X. Q., Yang, L. C., Xiong, F., Li, C. B., Li, F., & Zhou, G. Y. (2018). Aboveground biomass of the alpine shrub ecosystems in Three-River Source Region of the Tibetan Plateau. *Journal of Mountain Science, 15*, 357–363.

O'Mara, F. P. (2012). The role of grasslands in food security and climate change. *Annals of Botany, 110*, 1263–1270.

Qian, J. X., Yang, J., Sun, W. D., Zhao, L. L., Shi, L., Shi, H. T., Liao, L., Dang, C. Y., & Dou, Q. (2025). Soil Organic carbon estimation and transfer framework in agricultural areas based on spatiotemporal constraint strategy combined with active and passive remote sensing. *Remote Sensing, 17*.

Ramoelo, A., Cho, M. A., Mathieu, R., Madonsela, S., & Kerchove, R. V. D. (2015). Monitoring grass nutrients and biomass as indicators of rangeland quality and quantity using random forest modelling and WorldView-2 data. *International Journal of Applied of Applied Earth Observation and Geoinformation, 43*, 43–54.

Ren, J. Z., Liang, T. G., Lin, H. L., Feng, Q. S., Huang, X. D., Hou, F. J., Zou, D. F., & Wang, C. (2011). Study on grassland's responses to global climate change and its carbon sequestration potentials. *Acta Prataculturae Sinica*.

Scurlock, J., & Hall, D. O. (1998). The global carbon sink: A grassland perspective. *Global Change Biology, 4*, 229–233.

Scurlock, J. M. O., Johnson, K., & Olson, R. J. (2002). Estimating net primary productivity from grassland biomass dynamics measurements. *Global Change Biology, 8*, 736–753.

Shi, W., Yue, T., & Zong, W. (2017). High accuracy surface modeling method combined with auxiliary variables for soil mapping. In *Geoscience & Remote Sensing Symposium*.

Shi, Y. F. (2003). Discussion on the present climate change from warm dry to warm wet in northwest China. *Quaternary Sciences, 23*, 152–164.

Sun, X. F., Yue, T. X., & Wang, Q. (2013). High accuracy surface modeling of grassland aboveground biomass. *Journal of Remote Sensing, 17*, 1060–1076.

Tong, C., Wu, J., Yong, S., Yang, J., & Yong, W. (2004). A landscape-scale assessment of steppe degradation in the Xilin River Basin, Inner Mongolia, China. *Journal of Arid Environments, 59*, 133–149.

Vincenzi, S., Zucchetta, M., Franzoi, P., Pellizzato, M., Pranovi, F., De Leo, G. A., & Torricelli, P. (2011). Application of a Random Forest algorithm to predict spatial distribution of the potential yield of Ruditapes philippinarum in the Venice lagoon, Italy. *Ecological Modelling, 222*, 1471–1478.

Wadoux, A., Minasny, B., & McBratney, A.B. (2020). Machine learning for digital soil mapping: Applications, challenges and suggested solutions. *Earth-Science Reviews, 210*.

Wang, L. A., Zhou, X., Zhu, X., Dong, Z., & Guo, W. (2016). Estimation of biomass in wheat using random forest regression algorithm and remote sensing data. *Crop Journal, 4*, 212–219.
Wang, Y. Y., Wu, G. L., Deng, L., Tang, Z. S., Wang, K. B., Sun, W. Y., & Shangguan, Z. P. (2017). Prediction of aboveground grassland biomass on the Loess Plateau, China, using a random forest algorithm. *Scientific Reports, 7*, 1–10.
Xia, J. Z., Ma, M. N., Liang, T. G., Wu, C. Y., Yang, Y. H., Zhang, L., Zhang, Y. J., & Yuan, W. P. (2018). Estimates of grassland biomass and turnover time on the Tibetan Plateau. *Environmental Research Letters, 13*, 1–31.
Xie, Y. C., Sha, Z. Y., Yu, M., Bai, Y. F., & Zhang, L. (2009). A comparison of two models with Landsat data for estimating above ground grassland biomass in Inner Mongolia, China. *Ecological Modelling, 220*, 1810–1818.
Xiu, X. M., Zhou, S. F., Chen, Q., Meng, J. H., Dong, W. Q., Yang, G. B., & Li, X. S. (2019). Aboveground biomass estimation of provincial scattered grassland based on Google Earth Engine and machine learning. *Bulletin of Surveying and Mapping, 504*, 50–56.
Xu, B., Yang, X. C., Tao, W. G., Qin, Z. H., Liu, H. Q., Miao, J. M., & Bi, Y. Y. (2008). MODIS-based remote sensing monitoring of grass production in China. *International Journal of Remote Sensing, 29*, 5313–5327.
Yan, F., Wu, B., & Wang, Y. (2015). Estimating spatiotemporal patterns of aboveground biomass using Landsat TM and MODIS images in the Mu Us Sandy Land, China. *Agricultural & Forest Meteorology, 200*, 119–128.
Yang, S. X., Feng, Q. S., Liang, T. G., Liu, B. K., Zhang, W. J., & Xie, H. J. (2018). Modeling grassland above-ground biomass based on artificial neural network and remote sensing in the Three-River Headwaters Region. *Remote Sensing of Environment, 204*, 448–455.
Yuan, X., Li, L., Tian, X., Luo, G., & Chen, X. (2016). Estimation of above-ground biomass using MODIS satellite imagery of multiple land-cover types in China. *Remote Sensing Letters, 7*, 1141–1149.
Yue, T. X., Chen, C. F., & Li, B. L. (2010). An adaptive method of high accuracy surface modeling and its application to simulating elevation surfaces. *Transactions in GIS, 14*, 615–630.
Yue, T. X., Du, Z. P., Song, D. J., & Gong, Y. (2007). A new method of surface modeling and its application to DEM construction. *Geomorphology, 91*, 161–172.
Yue, T.X., Du, Z.P., & Liu, J.Y. (2004). High Accuracy Surface Modeling and Error Analysis. *Progress in Natural Science*, 62–68.
Yue, T. X., Liu, Y., Zhao, M. W., Du, Z. P., & Zhao, N. (2016). A fundamental theorem of Earth's surface modelling. *Environmental Earth Sciences, 75*, 751.
Zeng, N., Ren, X. L., He, H. L., Zhang, L., Li, P., Li, Z. Q., & Zhang, L. B. (2017). Aboveground biomass of Grasslands in the Three-River headwaters based on neural network. *Research of Environmental Sciences, 30*, 62–69.
Zhao, M., Yue, T., Na, Z., Sun, X., & Zhang, X. (2014). Combining LPJ-GUESS and HASM to simulate the spatial distribution of forest vegetation carbon stock in China. *Journal of Geographical Sciences, 24*, 249–268.
Zhao, N., Yue, T., Chen, C., Zhao, M., & Du, Z. (2018). An improved HASM method for dealing with large spatial data sets. *Science China Earth Sciences*, 1–10.
Zhao, Y. C., Wang, M. Y., Hu, S. J., Zhang, X. D., Ouyang, Z., Zhang, G. L., Huang, B. A., Zhao, S. W., Wu, J. S., Xie, D. T., Zhu, B., Yu, D. S., Pan, X. Z., Xu, S. X., & Shi, X. Z. (2018b). Economics- and policy-driven organic carbon input enhancement dominates soil organic carbon accumulation in Chinese croplands. *Proceedings of the National Academy of Sciences of the United States of America, 115*, 4045–4050.
Zheng, D. L., Rademacher, J., Chen, J. Q., Crow, T., Bresee, M., le Moine, J., & Ryu, S. R. (2004). Estimating aboveground biomass using Landsat 7 ETM+ data across a managed landscape in northern Wisconsin, USA. *Remote Sensing of Environment, 93*, 402–411.
Zheng, Y. R., Xie, Z. X., Robert, C., Jiang, L. H., & Shimizu, H. (2006). Did climate drive ecosystem change and induce desertification in Otindag sandy land, China over the past 40 years? *Journal of Arid Environments, 64*, 523–541.

Zhou, T., Geng, Y. J., Chen, J., Liu, M. M., & Lausch, A. (2020). Mapping soil organic carbon content using multi-source remote sensing variables in the Heihe River Basin in China. *Ecological Indicators, 114*, 106288.

Zhou, W., Yang, H., Huang, L., Chen, C., Lin, X., Hu, Z., & Li, J. (2017). Grassland degradation remote sensing monitoring and driving factors quantitative assessment in China from 1982 to 2010. *Ecological Indicators, 83*, 303–313.

Chapter 4
Hyperspectral Retrieval of Soil Nutrients

Abstract Soil nutrients are the main attributes of the soil, affecting the quality of the soil and the growth of plants. This study aims to use hyperspectral remote sensing data to construct an inversion model of soil organic carbon, total nitrogen, total phosphorus, and total potassium content in the source area of the Three Rivers and draws some conclusions and laws suitable for soil nutrients in the source area of the Three Rivers. It uses a variety of mathematical transformations to optimize the soil spectral data. It eliminates or reduces the influence of irrelevant noise, and improve the sensitivity of soil nutrients to the spectrum. It uses multiple linear regression (MLR), partial least square regression (PLS), support vector machine (SVM), and random forest (RF) to establish a hyperspectral inversion model of soil nutrients. Comparing the simulation results of the four models, it is found that the overall accuracy of RF is the highest. Among the four soil nutrient indexes, the multivariate scattering correction-first-order differential-random forest (MSCFD-RF) model has a higher prediction accuracy for soil organic carbon and total nitrogen, and the multivariate scattering correction-random forest (MSC-RF) model is more accurate The simulation results of potassium are better, and the R^2 of the verification set accuracy are 0.8739, 0.8592, and 0.8865 respectively, which shows that the prediction effect is relatively ideal. The verification set accuracy R^2 of all inversion models of soil total phosphorus is lower than 0.2, and the RPD value is lower than 1.4. It is impossible to predict the total phosphorus content, and the model fitting effect is not ideal.

Keywords Soil nutrient · Hyperspectral · Machine learning · Spectral transformation · Three-River headwaters region

4.1 Introduction

There are people in the earth, and everything is born in the earth. Soil nutrients are nutrients necessary for plant growth provided by the soil, and their composition and content have a decisive effect on soil quality. Therefore, obtaining the information in

© The Author(s), under exclusive license to Springer Nature Singapore Pte Ltd. 2026
W. Zhou et al., *Carbon Sink Estimation and Evaluation of Ecological Function Changes on the Tibetan Plateau*, https://doi.org/10.1007/978-981-95-8664-6_4

the soil quickly and accurately is of great significance for understanding soil monitoring and agricultural production. The soil spectrum is a comprehensive reflection of the chemical composition and structural characteristics of the internal material of the soil. Obtaining and interpreting soil spectra can effectively obtain a large amount of soil information. The rapid development of hyperspectral technology has increased the potential for quantitatively obtaining the content of soil chemical components (Nan, 2017).

The research on quantitative monitoring of soil nutrient elements has been a hot issue in the field of soil science research. In recent years, many scholars at home and abroad have conducted research on soil nutrient estimation based on hyperspectral data, extracting specific spectral feature indicators from hyperspectral data from multiple sources (Du, 2018). Nowadays, soil research based on hyperspectral remote sensing is mainly reflected in three aspects. One is the study of soil spectral reflectance characteristics including spectral curve feature absorption studies (Yang, 2014). The second is the study of the pre-processing methods of soil spectral reflectance data including soil spectral smoothing and denoising and spectral data transformation forms that affect the accuracy of inversion estimation (Nan, 2017). The third is the use of different inversion models for different types of soil types and the comparative analysis of accuracy (Yang, 2014).

With the development and maturity of hyperspectral technology, most of the studies on soil nutrient estimation by scholars at home and abroad in recent years have been based on hyperspectral data. Due to the diversity of soil constituent elements and the unique spectral characteristics of constituent elements, specific spectral feature indicators can be extracted from hyperspectral data from multiple sources (Du, 2018; Liu et al., 2011). Using the different spectral reflectance properties of soil samples, information on the quality of soils in the study area can be quickly obtained (Gao et al., 2013; Huang & Liu, 1995; Peng et al., 2006; Stenberg, 2010; Viscarra & Webster, 2011). Studies on the spectral reflectance of soils have been conducted earlier abroad. Ben-Dor et al. concluded that soil organic matter affects reflectance throughout the visible, near-infrared and short-wave infrared wavelength ranges, with the greatest effect in the visible range (Ben-Dor et al., 1997). Lee et al. showed that soil nitrogen absorption was best around 510 nm, but there were differences between different soil types (Suk Lee et al., 2001). Dalal et al. found that the soil total nitrogen content was closely related to the spectrum, roughly in the 1700–2100 nm band, and effectively estimated the total nitrogen content using the near-infrared spectral reflectance in the 1744, 1870 and 2052 nm bands (Dalal & Henry, 1986). Although research in soil spectral reflectance studies started relatively late in China, it is developing at a relatively fast pace (Qi et al., 2018). In the 1990s, Xu et al. studied soil spectral curves in China, and concluded that the content of organic matter in soil could be measured by the size of the "bow curve difference" at 600 nm (Xu & Dai, 1980). In their study of the moisture, organic matter, and total nitrogen content of soils, some research chose the yellow soil region of China and used NIR spectral analysis in their study to find the best combination of pulse points between NIR spectra and soil constituents (Peng et al., 1998).

Soil spectra are susceptible to a variety of influences including soil-forming matrix, grain size and physicochemical composition. Some scholars have used various pre-processing methods for soil spectral reflectance data to improve the accuracy of the model. Before building the model, some mathematical methods were applied to transform the soil spectrum, and the final model accuracy was better. Yu et al. measured the 400–1200 nm spectra of air-dried soils with particle sizes of 2 mm and 0.15 mm, respectively, and developed a linear model to estimate the content of total nitrogen and alkali-dissolved nitrogen. It was shown that the coefficient of determination of measured and predicted values for 0.15 mm grain size ($R^2 = 0.90$) was higher than that for 2 mm grain size ($R^2 = 0.88$) (Yu et al., 2002). In a study on the effect of soil grain size on modeling accuracy, Liu et al. used a partial least squares model for each of the five grain sizes of the Hubei tidal soil samples. The results showed to be influential (Liu et al., 2011). Before building the inverse model. Zhang et al. (2012) used two methods, SG smoothing convolution and Norris smoothing filtering, respectively, for the pre-processing of smoothing and noise reduction on the basis of the first-order inverse spectra. It was concluded that these two methods could reduce the background noise while also improving the accuracy of the inverse model.

Meanwhile, along with the advancement of computer technology, nonlinear regression models such as artificial neural networks, random forests, and support vector machines have been gradually applied to the study of soil organic matter prediction. The nonlinear models have strong linear and nonlinear processing capabilities, and by selecting different parameters to construct integrated models, the predictive performance of the models can be further improved. A number of related studies have used stepwise regression models to simulate the prediction of soil organic matter in the semi-arid zone of Israel and the southern and eastern regions of Africa, and the accuracy of the inverse models has performed well. A combination of principal component regression models with first-order derivatives was used to model soil samples within a region of the USA, and the inversion results achieved a correlation coefficient of 0.87 (Chang & Laird, 2002). In China, some scholars have used different transformed forms of hyperspectral reflectance with multiple stepwise regression models in the exploration of soil organic matter, and the prediction accuracy achieved was also satisfactory (Peng et al., 2006). Ji et al. (2012) compared several nonlinear models such as random forest models, support vector machine models and artificial neural networks in order to better invert the organic matter content of soil samples from rice fields in Zhejiang Province. The modeling accuracy of the nonlinear models was found to be relatively higher than that of the partial least squares regression models, with the support vector machine model being the most prominent among the models. Later, artificial neural network methods were introduced into the traditional partial least squares regression model, and the prediction accuracy of the model was significantly better than that of the direct partial least squares regression model.

The diversity of soil nutrients makes it impossible to find an optimal model for the inversion of the content of all soil nutrient indicators. The results of (Yuan et al., 2009) provide good evidence that the partial least squares regression model fits better than the support vector machine model in the study of total soil nitrogen content. However, in the study of total phosphorus content it is exactly to the left. At present, studies using hyperspectral inversion of soil properties are mainly about moisture, soil organic matter, heavy metals and indicators such as phosphorus, nitrogen, and potassium with soil reflectance spectra. The inversion results are also relatively satisfactory. The above studies fully demonstrate the feasibility of hyperspectral techniques in the inversion of soil nutrients, and provide a theoretical basis and reference for the study of this paper (Qi et al., 2018).

4.2 Materials and Methods

4.2.1 Study Area

The study area is the Jiang Yuan District (89° 45′–102° 23′ E, 31° 39′–36° 12′ N), which is located in the south of Qinghai Province at an average altitude of 3500–4800 m. The total land area is 302,500 km^2, accounting for about 43% of the total area of Qinghai Province. It is known as the "Chinese Water Tower" because it is the birthplace of the Yangtze, Yellow, and Lancang Rivers. The Three-rivers source region (TRSR) region is the hinterland of the Tibetan Plateau and is dominated by mountainous landscapes, with high terrain and complex topography. The region is covered with snow-capped glaciers and rivers, and the total area of wetlands is 73,300 km^2, making it the highest natural wetland in China. The TRSR region has a typical plateau continental climate, with alternating hot and cold seasons, and distinct dry and wet seasons. It has a unique alpine ecosystem, typical of the alpine environment of the Central Asian plateau and the alpine grasslands of the world. The region is rich in flora and fauna and blessed with tourism resources (Fig. 4.1).

4.2.2 Soil Type

The soils of the TRSR area belong to the mountainous soil system of the Tibetan Plateau. Due to the young age of geological development on the Tibetan Plateau, the soil matrix is relatively primitive and the soil formation time is short. As a result, most of the soils in the area are thin, coarse, with poor water retention capacity and low fertility, and are susceptible to erosion and soil erosion. Due to the large difference in altitude within the region, the soil types in the TRSR area show a vertical zonal distribution pattern. As the altitude goes from high to low, the soil types are mainly alpine cold desert soil, alpine meadow soil, alpine grassland soil, mountain meadow soil, gray-brown soil, chestnut calcium soil, and mountain forest soil, with alpine meadow soil being the main type (Fig. 4.2). The differences in texture and physical

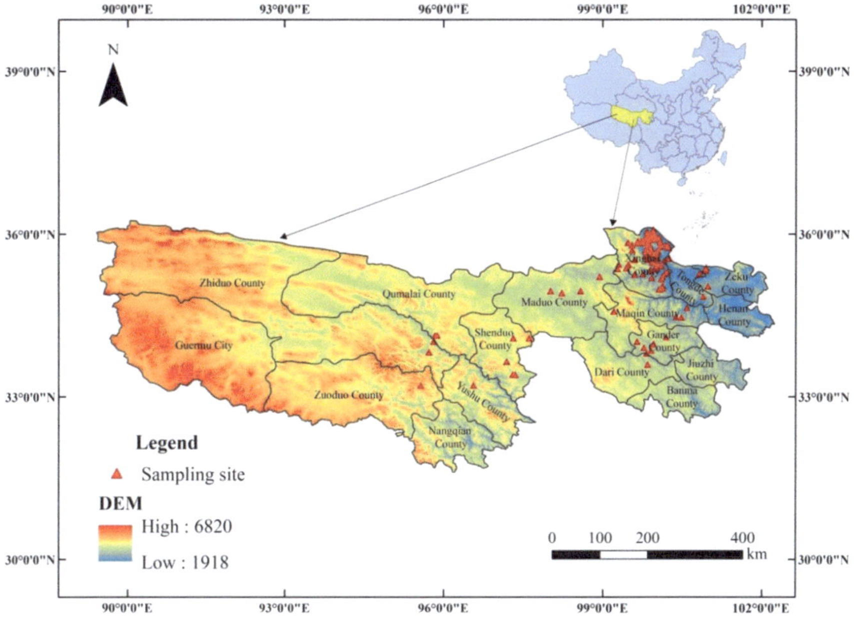

Fig. 4.1 Geographical location of TRSR and distribution of sampling sites

properties between soil types are one of the influencing factors for the spatial variation in soil nutrient content in the TRSR area.

4.2.3 *Soil Sample Collection and Processing*

4.2.3.1 Sample Collection

The soil samples used in this study were taken from the TRSR area and sampled during the months of July to August in 2017, 2018, 2019, and 2020. The low precipitation and low soil moisture in the TRSR area during this time period facilitated later soil air-drying. A total of 294 soil samples were collected (272 after excluding abnormal samples), taking into account soil type, vegetation type, accessibility, and scientific sampling. Soil sampling sites were mainly located in Xinghai, Mado, Machin, and Gander counties, and to a lesser extent in Weiduo, Zhiduo, Zhaoduo, and Yushu counties (Fig. 4.1). Topsoil samples of 0–30 cm and about 1 kg were collected, mixed evenly and sealed in a dense bag, and brought back well marked (Yang, 2014). At the same time, the latitude and longitude of the sampling points were recorded using the OWI interactive map or handheld GPS, and the information related to the soil vegetation type of the sampling points was recorded and photographed.

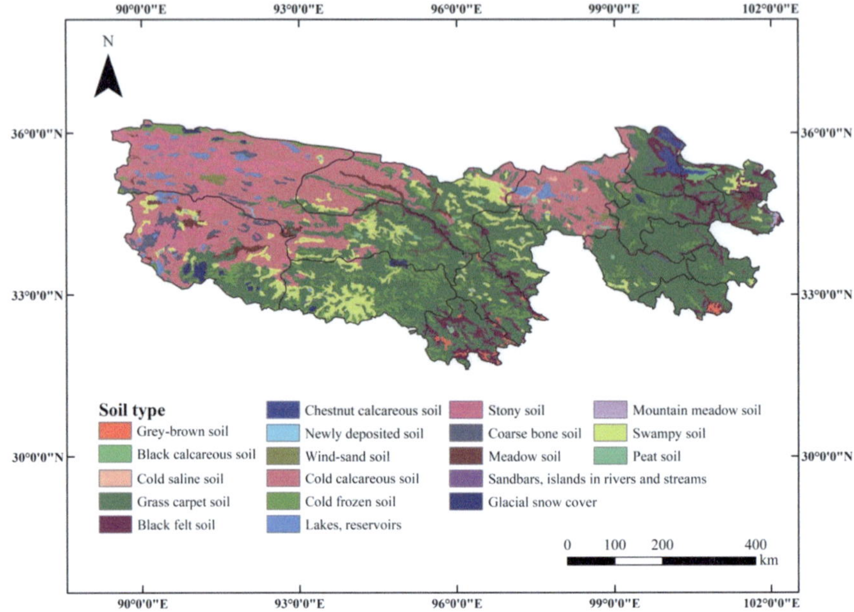

Fig. 4.2 Map of the distribution of soil types in the TRSR area

4.2.3.2 Sample Processing

The collected soil samples were placed indoors for natural air-drying, with the aim of reducing the moisture in the soil. Next, plant and animal debris and stones were removed from the soil and passed through 10-, 60- and 100-mesh soil sieves to obtain samples of 2 mm, 0.25 mm, and 0.15 mm, respectively (Du Rongxiang, 2018). Finally, the treated soil samples were divided into three parts, two of which were used for indoor chemical analysis and spectral collection, and the other part was sealed for investigation (Du, 2018).

4.2.3.3 Determination of Physical and Chemical Properties of the Soil

Soil samples of 0.25 mm were taken for the determination of soil physicochemical indicators at the Chongqing Key Laboratory of Karst Environment, Southwest University. In this study, the organic carbon (SOC), total nitrogen (TN), total phosphorus (TP) and total potassium (TK) contents of 272 soil samples were determined. Descriptive statistics were used to reflect the statistical characteristics of soil nutrient element contents in the TRSR area. In descriptive statistics, the mean value is usually used to reflect the concentration trend of the sample data. The standard deviation and coefficient of variation are used to reflect discrete trends in the sample data. The skewness and kurtosis coefficients are used to reflect the distribution pattern of the

Table 4.1 Chemical methods for the determination of soil nutrients

Samples	Determination method
SOC	Potassium dichromate external heating method
TN	Kjeldahl distillation
TP	Molybdenum antimony anti-colorimetric method
TK	Alkali fusion—flame photometry

Table 4.2 Statistical table of soil nutrient results for valid soil samples

Samples	Minimum (g/kg)	Maximum (g/kg)	Mean (g/kg)	Standard deviation (SD)	Kurtosis	Skewness	Coefficients of variation/%	Soil nutrient classes
SOC	2.72	131.44	34.68	22.92	3.20	1.64	66.10	1(≥ 23.2)
TN	0.12	10.32	2.30	2.16	0.91	1.11	93.91	1(≥ 2)
TP	0.24	1.48	0.72	0.21	0.78	0.50	28.83	2(0.8–1)
TK	7.84	33.19	19.87	7.36	− 1.50	0.50	37.04	3(15–20)

sample data (Du, 2018). The methods and statistical results for the determination of the components of the soil in the TRSR area are shown in Tables 4.1 and 4.2.

4.2.3.4 Soil Spectral Data Determination

An ASD FieldSpec4 portable geophysical spectrometer was used to determine the indoor spectra of soil samples in a dark room without light (Yang, 2014). Its spectral detection range was from 350 to 2500 nm, with a final total of 2151 bands output (Fig. 4.3). Next, a total of 20 data were collected in four directions during the spectral determination. Finally, after removing the outliers, the average value was used as the standard spectral reflectance for each sample (Yang, 2014). During this period, a whiteboard calibration is generally performed once in about 10 min in order to reduce the errors introduced by the instrument during the test.

4.2.3.5 Spectral Transformation Methods

Soil spectra are susceptible to varying degrees of influence from the measurement instrument, the measurement environment and the roughness of the soil sample in the process of measurement. This makes the signal-to-noise ratio of soil spectra low and the relationship between spectral properties and constituent content difficult to establish accurately. At the same time, a single raw spectral reflectance cannot reveal the spectral reflectance characteristics of all nutrient indicators. Multiple mathematical transformation methods for different nutrient indicators can help improve the sensitivity of soil nutrients to spectra and eliminate or attenuate the influence of extraneous

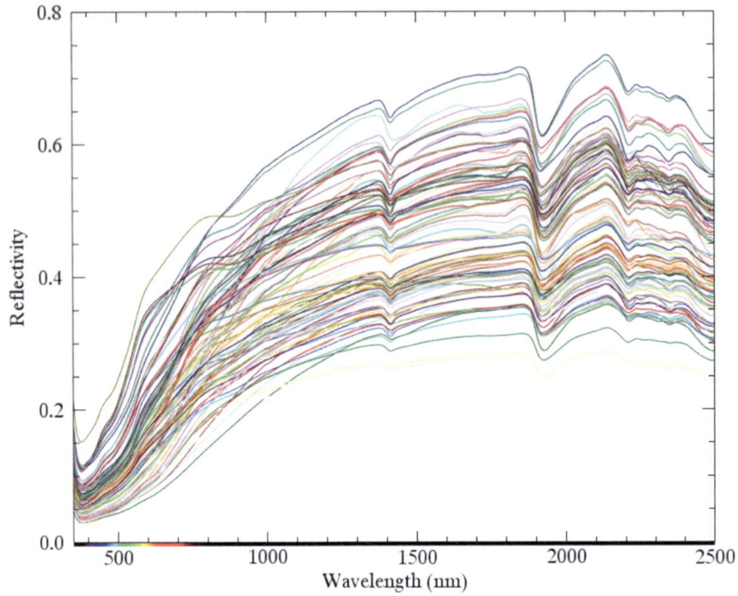

Fig. 4.3 Raw spectral profiles of all soil samples in the TRSR area

noise. Combined with previous research results, the five main spectral transformation methods used in this paper are first-order differentiation (FD), second-order differentiation (SD), logarithm of the inverse (LR), removal of the envelope (CR), and multiple scattering correction (MSC), as well as the completion of first-order differentiation (FD) based on removal of the envelope (CR) and multiple scattering correction (MSC) for subsequent modeling.

The differential transformation of a spectrum is a common method of spectral transformation by calculating differential values of different orders to reveal some information hidden in the spectral curve. Some studies have shown that in the process of spectral transformation, the lower order differential is less sensitive to the influence of noise and works better in the practical application of soil composition inversion. On the contrary differential transformations of too high an order tend to eliminate some effective information in the soil (Yang, 2014). Therefore, FD is usually used to eliminate the influence of partially linear or near-linear background and noise spectra on the target spectrum, to improve spectral resolution and sensitivity, and to easily find bands with high correlation (Yu et al., 2015). SD is sufficient to eliminate the effects caused by baseline drift and sample size (Yu et al., 2015). First-order differentiation and second-order differentiation are defined in Eq. (4.2).

$$R'(\lambda_i) = \left[R(\lambda_{i+1}) - R(\lambda_{i-1})\right]/(\lambda_{i+1} - \lambda_{i-1}) \tag{4.1}$$

$$R''(\lambda_i) = \left[R'(\lambda_{i+1}) - R'(\lambda_{i-1})\right]/(\lambda_{i+1} - \lambda_{i-1}), \tag{4.2}$$

where λ_{i-1}, $\lambda_{i,}$ and λ_{i+1} denotes adjacent wavelengths (nm). $R(\lambda_{i+1})$ denotes the wavelength λ_{i+1} corresponding to the original spectral reflectance. $R'(\lambda_i)$ denotes the first-order differentiation of the original spectrum corresponding to wavelength λ_i. $R''(\lambda_i)$ denotes the second-order differentiation of the original spectrum corresponding to wavelength λ_i.

The log–log transformation (LR) of soil spectral reflectance can be used to improve spectral differences in the visible region and to reduce the effects of multiplicative factors caused by changes in light conditions and topography (Lu et al., 2007).

$$R'(\lambda_i) = \log[1/R(\lambda_i)], \tag{4.3}$$

where $R'(\lambda_i)$ represents the spectral reflectance value after processing. $R(\lambda i)$ is the original spectral reflectance on band λi.

The d-envelope, also known as continuum removal (CR), is one of the commonly used spectral transformation methods. The essence of the method is to remove the reflectance value of the corresponding band on the envelope with the measured spectral reflectance value. The reflectance spectrum is normalized to between 0 and 1 (Yang, 2014). It reduces the influence of background information, expands the absorption and reflection properties of the spectral profile and classifies and identifies the extracted sensitive wavebands (Du, 2018). The formula for the de-envelope calculation is given in formula 4.4.

$$R'(\lambda) = R(\lambda)/R_C(\lambda), \tag{4.4}$$

where $R'(\lambda)$ denotes the spectral reflectance value after processing. $R(\lambda)$ is the original spectral reflectance of a band. $R_c(\lambda)$ corresponding to the reflectance value on the continuum on the band.

The principle of the multiple scattering correction (MSC) transformation is to separate the spectral scattering signal from the chemical absorption information. The method effectively removes the effect of scattering, improves the spectral absorption information related to the component content and further improves the signal-to-noise ratio of the spectrum (Zhang et al., 2018a, 2018b). The effect of the multiple scattering correction of the reflectance of the soil spectrum of TRSR is shown in Fig. 3.5. The formulae for the multiple scattering correction are given in Eqs. (4.5) and (4.6).

$$R_i = m_i\overline{R_{i,j}} + b_i \tag{4.5}$$

$$R_i' = [R_i - b_i]/m_i, \tag{4.6}$$

where m_i and b_i are the relative offset coefficients and parity of each spectrum after a one-dimensional linear regression with the mean spectrum, respectively (Zhang et al., 2018a, 2018b). $\overline{R_{i,j}}$ is the mean value of each band of each spectrum.

4.2.4 Prediction Models

We applied four algorithms (i.e., MLR, PLS, SVM, and RF models) to rapidly determine the soil nutrient content in the TRSR region. These models are able to simulate complex linear or nonlinear relationships between soil nutrients and hyperspectral data.

Multiple Linear Regression (MLR) makes predictions or estimates from the best combination of multiple independent variables. In performing hyperspectral modeling, MLR can improve the fit to a partial extent by increasing the number of variables. At the same time, the method can create optimal multiple linear regression equations by analyzing and testing the significance of the linear effects of multiple independent variables on the dependent variable and screening those with significant effects. This method is good for solving the problems of high number of hyperspectral data bands and the existence of information redundancy (Zhang et al., 2018a, 2018b).

The Partial Least Squares (PLS) method finds the best matching function for a data set by minimizing the sum of squares of the errors (Yang, 2014). This method can be combined with the advantages of multiple linear regression analysis (MLR), canonical correlation analysis (CCA), and principal component analysis (PCA) (Yang, 2014) to perform some of the similar studies as principal component analysis and canonical correlation analysis and to provide richer and deeper information (Yang, 2014). Secondly, the spectral dimensionality can be reduced and the main controlling factors influencing the variation in nutrient content can be identified. Third, it enhances the interpretation of spectral data and improves the accuracy of the analysis results (Zhang et al., 2018a, 2018b).

Support Vector Machine (SVM) is a two classification model (Yang, 2014), the principle of which is to find the farthest distance between several sample points and the hyperplane. In other words the method is that to the maximum interval hyperplane. Support vector machines are suitable for problems with a small number of samples, linear and nonlinear and high feature dimensions because of their simple structure, adaptability, generalization and generalization capabilities. It is not prone to over-fitting.

Random Forest (RF) is a type of integrated learning algorithm. It integrates multiple decision trees to make predictions and is a method for analyzing nonlinear relationships between complex variables. The Random Forest method is widely used in nonlinear, high-dimensional data problems. It can prevent accuracy degradation due to noisy training samples and missing data, while also improving computational efficiency, and is quite good at solving both classification and regression types of problems.

4.2.5 Model Validation

Different combinations of spectral transformations have been constructed to explore the rapid determination of soil nutrient content in the TRSR region based on soil nutrients and indoor hyperspectral data. We then compared the predictive performance of the different combinations based on MLR, PLS, SVM, and RF methods. These models were compared and evaluated using the K-fold cross-validation method, with the full sample 272 in a 2:1 ratio to create a modeling set 182 and a validation set 90. The multi-species soil nutrient inversion models developed are subject to accuracy validation. Model accuracy measures the overall estimation capability of a model and evaluates the fit between measured and predicted values. The better the modeling accuracy, the better the stability of the model, and the higher the validation accuracy, the better the predictive ability of the model (Lu et al., 2007). In this paper, three indicators, namely Coefficient of Determination (R^2), Root-Mean-Square error (RMSE), and Relative Predictive Deviation (RPD), are selected as indicators for evaluating the model fit superiority. These validation metrics are defined as shown in Eqs. (4.7)–(4.9).

$$R^2 = 1 - \frac{\sum_{i=1}^{n} (\hat{y}_i - y_i)^2}{\sum_{i=1}^{n} (y_i - \overline{y})^2} \tag{4.7}$$

$$\text{RMSE} = \sqrt{\frac{1}{n}\sum_{i=1}^{n}(\hat{y}_i - y_i)^2} \tag{4.8}$$

$$\text{RMSE} = \sqrt{\frac{\sum_{i=1}^{n} (\hat{y}_i - y_i)^2}{n}} \tag{4.8}$$

$$\text{RPD} = \frac{\text{SD}}{\text{RMSE}} = \frac{\sqrt{\frac{1}{n}\sum_{i=1}^{n} (y_i - \overline{y})^2}}{\sqrt{\frac{1}{n}\sum_{i=1}^{n} (\hat{y}_i - y_i)^2}} = \sqrt{\frac{\sum_{i=1}^{n} (y_i - \overline{y})^2}{\sum_{i=1}^{n} (\hat{y}_i - y_i)^2}}, \tag{4.9}$$

where y_i is the i soil elemental content in the validation set. $\hat{y}_i$ is the predicted value of the i soil elemental content. $\bar{y}$ is the mean of the measured soil nutrient element content. And n is the number of samples in the validation set.

4.3 Results

4.3.1 Descriptive Statistics

The statistical results for the SOC, TN, TP, and TK contents of the soil samples and the values of the corresponding predictor variables are shown in Table 4.2 and Fig. 4.4. SOC content in the table ranged from 2.72 to 131.44 g/kg, with an SD of 22.92 g/kg. TN content ranged from 0.12 to 10.32 g/kg, with an SD of 2.16 g/kg. TP content was relatively low, ranging from 0.24 to 1.48 g/kg with an SD of 0.21 g/kg. TK content was relatively abundant (mean is 19.87 g/kg) with a standard deviation of 7.36 g/kg. SD values for SOC, TN, TP, and TK content were all below the mean, indicating a large variation between samples and moderate variation in their distribution.

The skewness and kurtosis coefficients of the samples were close to 0. Combining Table 4.2 and Fig. 4.4, some conclusions can be drawn. The skewness of both SOC and TN was greater than 1, indicating that their contents were heavily skewed. The distribution of SOC and TN contents was uneven, with more distribution of low contents and less distribution of high contents. The skewness of TP and TK was 0.5, indicating that they were slightly skewed. The distribution of TK content was bimodal, with clear boundaries between high and low values. According to Nielsen criteria, the coefficients of variation of the four soil nutrients in the TRSR area, SOC, TN, TP, and TK, are all moderately variable (0.1 < CV < 1).

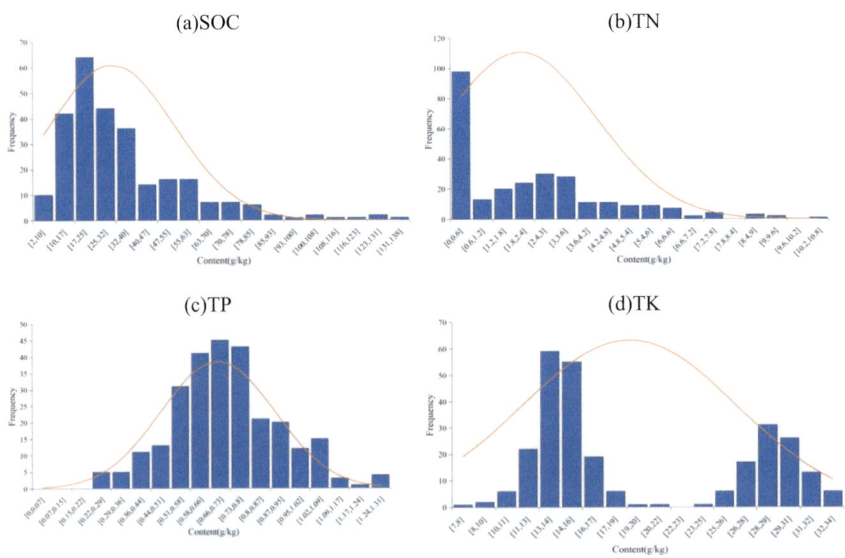

Fig. 4.4 Soil nutrient content frequency map

4.3.2 *Spatial Distribution Characteristics*

We explore the spatial distribution characteristics of soil nutrients in the TRSR area from both horizontal and vertical perspectives, exploring the influence of vegetation type and altitude on the distribution of soil nutrient content.

4.3.2.1 Vegetation Type

Tables 4.3, 4.4, 4.5 and 4.6 show the results of the statistical analysis of soil nutrients according to vegetation type. From Table 4.3 it can be seen that the average SOC content is higher in the marsh grassland and swampy meadow types, and less in the desert grassland type. This indicates that SOC content is positively correlated with vegetation cover, height and microbial load. The coefficient of variation of SOC content was higher in vegetation types such as alpine meadows. This may be related to vegetation distribution and exposure to rodent infestation when analyzed in relation to actual conditions. From Table 4.4 it can be seen that the average TN content is higher in swampy grassland and swampy meadows. It was also found that the pattern of TN content was consistent with that of SOC. Soil total nitrogen content was higher in types with higher moisture content and higher vegetation cover, and lower in areas with uneven vegetation distribution and high levels of rodent infestation.

Analysis of Table 4.5 shows that TP levels are low in all vegetation types and the distribution pattern is similar to that of SOC and TN. The coefficient of variation did not differ significantly across all vegetation types, indicating that the amount of total phosphorus available in the soil was less correlated with vegetation type. This may be related to the fact that most of it is not directly absorbed by plants in the same way. From Table 4.6 it is concluded that the average TK content is higher in the woodland and farmland types. The main reason for this may be that potassium is in a mobile state in the soil and where there is more soil moisture, the uptake of potassium by vegetation is more favorable. The higher levels of potassium in the farmland are mainly due to fertilizer application.

4.3.2.2 Altitude

All the sampling points were statistically analyzed and it was found that the study sample points were located between 2500 and 5000. Soil nutrient content showed different patterns with altitude. Tables 4.7, 4.8, 4.9 and 4.10 shows the results of the statistical analysis of soil nutrients according to poster altitude. The analysis of Tables 4.7, 4.8 and 4.9 shows that the distribution patterns of SOC, TN, and TP were consistent. The SD values of SOC, TN, TP, and TK contents were all lower than the mean values, indicating a large variation between samples. The mean contents of SOC, TN, and TP increased with increasing altitude, and the mean content of TK increased and then decreased.

Table 4.3 Results of SOC content of different vegetation types in the TRSR area

Samples	Type of vegetation	Number of samples	Minimum (g/kg)	Maximum (g/kg)	Mean (g/kg)	Standard deviation (SD)	Kurtosis	Skewness	Coefficients of variation (%)
SOC	Alpine meadow	61	14.49	126.14	48.59	23.26	1.55	1.12	47.87
	Alpine grassland	59	10.17	71.08	27.89	14.66	0.38	1.05	52.55
	Alpine meadow grassland	44	11.72	108.71	37.45	21.37	1.92	1.44	57.08
	Temperate grassland	43	13.80	102.11	26.46	14.04	17.86	3.98	53.04
	Desert grassland	24	2.72	26.48	15.38	7.87	− 1.24	− 0.33	51.20
	Farmland	13	10.78	23.41	18.15	4.01	− 0.98	− 0.63	22.12
	Mountain meadows	12	28.79	82.80	51.33	17.08	− 1.17	0.37	33.27
	Swampy meadows	5	27.28	103.83	69.23	28.02	− 1.16	− 0.36	40.47
	Woodland	3	38.33	58.90	45.52	11.60	− 2.00	0.57	25.48
	Temperate desert grassland	3	15.00	19.60	17.30	2.30	− 2.00	0.01	13.28
	Swampy grassland	2	128.47	131.44	129.96	2.09	− 2.50	0.00	1.61

Table 4.4 Results of TN content of different vegetation types in the TRSR area

Samples	Type of vegetation	Number of samples	Minimum (g/kg)	Maximum (g/kg)	Mean (g/kg)	Standard deviation (SD)	Kurtosis	Skewness	Coefficients of variation (%)
TN	Alpine meadow	61	0.18	10.32	3.61	2.65	− 0.57	0.41	73.30
	Alpine grassland	59	0.13	5.47	1.62	1.45	− 0.19	0.87	89.20
	Alpine meadow grassland	44	0.31	8.52	3.16	1.98	0.12	0.72	62.57
	Temperate grassland	43	0.16	3.70	1.66	1.27	− 1.76	− 0.03	76.79
	Desert grassland	24	0.13	3.02	0.99	0.96	− 0.99	0.80	96.66
	Farmland	13	0.12	0.25	0.20	0.04	− 1.06	− 0.47	21.59
	Mountain meadows	12	0.39	4.93	2.27	1.82	− 1.73	0.27	80.35
	Swampy meadows	5	2.32	7.53	5.36	1.96	− 1.17	− 0.58	36.60
	Woodland	3	0.34	0.45	0.39	0.06	− 2.00	0.19	14.60
	Temperate desert grassland	3	0.16	0.20	0.19	0.02	− 2.00	− 0.49	11.55
	Swampy grassland	2	4.51	9.21	6.86	3.32	− 2.50	0.00	48.43

Table 4.5 Results of TP content of different vegetation types in the TRSR area

Samples	Type of vegetation	Number of samples	Minimum (g/kg)	Maximum (g/kg)	Mean (g/kg)	Standard deviation (SD)	Kurtosis	Skewness	Coefficients of variation (%)
TP	Alpine meadow	61	0.33	1.08	0.70	0.17	− 0.42	0.25	24.72
	Alpine grassland	59	0.29	1.12	0.70	0.20	− 0.49	− 0.19	28.14
	Alpine meadow grassland	44	0.24	1.09	0.68	0.22	− 0.61	0.29	32.55
	Temperate grassland	43	0.49	1.48	0.82	0.24	0.63	1.15	29.04
	Desert grassland	24	0.29	0.83	0.61	0.17	− 0.83	− 0.60	28.06
	Farmland	13	0.67	1.27	0.84	0.16	1.51	1.42	19.34
	Mountain meadows	12	0.56	1.20	0.84	0.18	− 0.74	0.52	21.74
	Swampy meadows	5	0.57	1.25	0.83	0.31	− 1.87	0.43	37.74
	Woodland	3	0.60	0.76	0.68	0.08	− 2.00	0.01	11.95
	Temperate desert grassland	3	0.62	0.74	0.68	0.06	− 2.00	0.21	8.61
	Swampy grassland	2	0.75	0.85	0.80	0.07	− 2.50	0.00	8.34

Table 4.6 Results of TK content of different vegetation types in the TRSR area

Samples	Type of vegetation	Number of samples	Minimum (g/kg)	Maximum (g/kg)	Mean (g/kg)	Standard deviation (SD)	Kurtosis	Skewness	Coefficients of variation (%)
TK	Alpine meadow	61	9.35	31.07	18.06	6.40	– 0.82	0.88	35.43
	Alpine grassland	59	11.20	32.43	20.37	7.25	– 1.73	0.29	35.60
	Alpine meadow grassland	44	11.40	30.94	16.33	5.40	1.89	1.85	33.10
	Temperate grassland	43	13.83	33.19	22.12	7.70	– 1.83	0.26	34.82
	Desert grassland	24	12.93	31.69	19.95	7.36	– 1.67	0.52	36.92
	Farmland	13	27.53	32.38	29.90	1.81	– 1.56	0.11	6.07
	Mountain meadows	12	13.82	30.00	21.73	6.88	– 2.02	0.01	31.65
	Swampy meadows	5	10.99	14.53	12.47	1.58	– 1.87	0.32	12.68
	Woodland	3	29.85	31.41	30.59	0.78	– 2.00	0.15	2.55
	Temperate desert grassland	3	28.52	30.59	29.82	1.14	– 2.00	– 0.55	3.81
	Swampy grassland	2	7.84	12.33	10.08	3.17	– 2.50	0.00	31.48

Table 4.7 Results of SOC content at different altitudes in the TRSR area

Samples	Altitude	Number of samples	Minimum (g/kg)	Maximum (g/kg)	Mean (g/kg)	Standard deviation (SD)	Kurtosis	Skewness	Coefficients of variation (%)
TN	<3000	5	2.77	11.78	7.23	3.50	− 1.57	− 0.01	48.42
	[3000, 3500)	64	11.81	60.57	25.22	11.22	1.47	1.44	44.49
	[3500, 4000)	32	11.77	82.80	36.05	15.98	0.48	0.68	44.33
	[4000, 4500)	38	10.17	93.63	39.96	22.68	− 0.52	0.59	56.75
	≥ 4500	8	54.15	128.47	87.53	32.04	− 1.81	0.32	36.60

Table 4.8 Results of TN content at different altitudes in the TRSR area

Samples	Altitude	Number of samples	Minimum (g/kg)	Maximum (g/kg)	Mean (g/kg)	Standard deviation (SD)	Kurtosis	Skewness	Coefficients of variation (%)
TN	<3000	5	0.12	0.95	0.56	0.35	− 1.80	− 0.06	62.86
	[3000, 3500)	64	0.16	6.50	1.81	1.37	0.87	0.75	75.73
	[3500, 4000)	32	0.18	4.83	2.04	1.62	− 1.52	0.21	79.46
	[4000, 4500)	38	0.13	8.62	3.53	1.98	− 0.26	0.45	56.21
	≥ 4500	8	4.51	10.32	6.58	2.18	− 1.06	0.81	33.11

Table 4.9 Results of TP content at different altitudes in the TRSR area

Samples	Altitude	Number of samples	Minimum (g/kg)	Maximum (g/kg)	Mean (g/kg)	Standard deviation (SD)	Kurtosis	Skewness	Coefficients of variation (%)
TP	<3000	5	0.29	0.97	0.49	0.28	− 0.72	1.07	57.54
	[3000, 3500)	64	0.42	1.48	0.79	0.23	0.55	0.99	29.62
	[3500, 4000)	32	0.37	1.12	0.73	0.21	− 1.04	0.08	28.91
	[4000, 4500)	38	0.24	1.06	0.62	0.21	− 0.33	0.14	34.55
	≥ 4500	8	0.57	1.06	0.75	0.17	− 0.88	0.67	22.13

Table 4.10 Results of TK content at different altitudes in the TRSR area

Samples	Altitude	Number of samples	Minimum (g/kg)	Maximum (g/kg)	Mean (g/kg)	Standard deviation (SD)	Kurtosis	Skewness	Coefficients of variation (%)
SOC	<3000	5	13.24	32.36	17.58	8.28	− 0.42	1.33	47.09
	[3000, 3500)	64	10.99	33.19	20.01	7.42	− 1.45	0.62	37.11
	[3500, 4000)	32	12.69	30.60	20.46	6.76	− 1.75	0.33	33.05
	[4000, 4500)	38	11.14	32.43	15.21	3.90	9.11	2.81	25.62
	≥ 4500	8	7.84	16.08	12.94	2.77	− 0.85	− 0.53	21.40

Analysis of Table 4.7 shows that the SD of SOC increased gradually with increasing altitude. The coefficient of variation of SOC content gradually decreases when the altitude is below 4000 m. The coefficient of variation of SOC content increased when the altitude was higher than 4000 m. This indicates that the variation in SOC content is relatively small and evenly distributed within the low altitude region. At higher altitudes, the variation in SOC content increases sharply and the spatial distribution is more discrete. Analysis of Table 4.8 shows that the SD of TN content increases gradually with increasing altitude. When the altitude is below 4000 m, the coefficient of variation of TN content gradually increases. When the altitude is higher than 4000 m, the coefficient of variation of TN content decreases. This indicates that the variation in TN content is relatively small and the distribution is more discrete within the low altitude region. This may be related to the changes in biomass due to human grazing activities. At higher altitudes, the variation in TN content increases sharply, but the spatial distribution is more concentrated.

Analysis of Table 4.9 shows that the SD and coefficient of variation of TP content gradually decrease with increasing altitude. This indicates that the variation in TP content in the TRSR area is small, and the coefficient of variation increases abruptly between 4000 and 4500. The reason for this variation may be related to the degree of rodent infestation. Soil erosion is more serious and there is more nutrient loss from the soil. Analysis of Table 4.10 shows that the SD and coefficient of variation of TK content gradually decrease with increasing altitude. This indicates that the distribution of TP content in the TRSR area is relatively concentrated with little variability. When the altitude was below 4000 m, the mean value of TK content gradually increased. When the altitude was higher than 4500 m, the mean value of TK content suddenly decreased. This is mainly due to the fact that as the altitude rises, the vegetation cover decreases and the distribution range is more concentrated, so the potassium content decreases and is concentrated.

4.3.2.3 Spatial Distribution Patterns

The soil nutrient contents of all the sampling sites in the TRSR area were kriged in ArcGIS. The difference results (Fig. 4.5) show that the distribution of soil organic carbon and total nitrogen in the TRSR area is generally consistent. The low values are distributed in the north, the high values in the middle and low altitude areas in the south. The medium values in the high altitude areas in the west. It shows a pattern of gradual decrease from the southeast and northwest inwards (Zhang et al., 2018a, 2018b). The distribution pattern of total phosphorus content is opposite to that of organic carbon and total nitrogen (Zhang et al., 2018a, 2018b). Low values are found at high altitudes in the west and high values are distributed in the north-eastern and southern regions. The distribution pattern of total potassium, on the other hand, shows an increase from all around to the north-east, with higher levels in the north-east. However, the pattern of variation along a particular direction is not strong (Zhang et al., 2018a, 2018b).

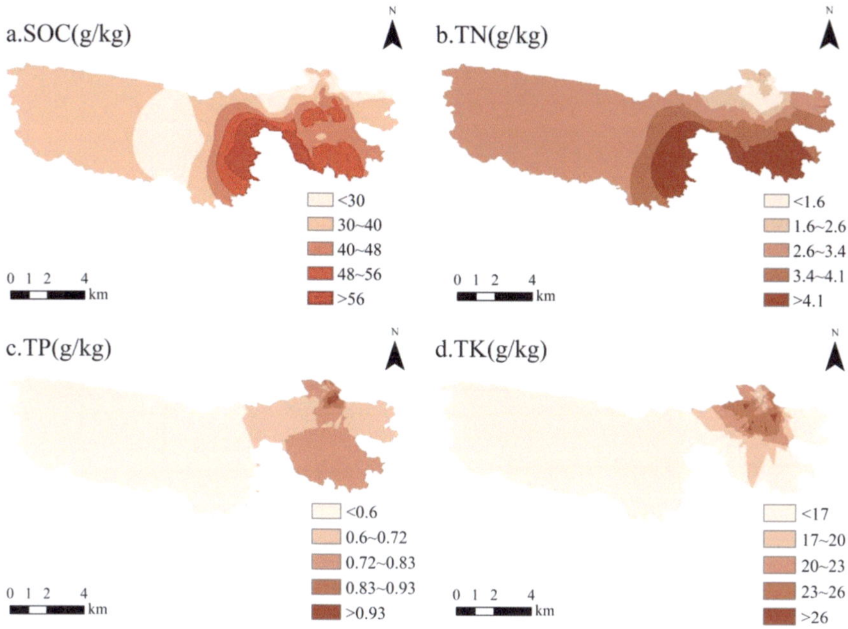

Fig. 4.5 Spatial distribution of soil nutrient content

4.3.3 Spectral Characterization and Feature Band Extraction

4.3.3.1 Soil Raw Reflectance Spectral Characterization

Soil samples were measured in the laboratory to obtain spectral reflectance and then the spectral data were smoothed in The Unscrambler X 10.4 using the moving average method. Figure 4.6 shows the results of plotting the raw spectral reflectance profile of the soil using ENVI software. Although there is some variation in the magnitude of the spectral reflectance of the different soil types, it can be observed that these soil spectral curves have some common features (Nan, 2017).

As can be seen from Fig. 4.6, there is consistency in the pattern, trend and location of the major peaks and troughs of the spectral reflectance curves for all soil samples. The spectral reflectance values are in the range of 0 to 0.8 and the spectral curves show dispersion in the 400 to 2450 nm band. It mainly contains the following manifestations. In the 400 to 1000 nm band range, the spectral curve tends to increase with increasing wavelength as the soil reflectance value increases significantly, with a gradual increase in slope and a steep curve trend. In the 1000–2200 nm band, the growth rate of soil spectral reflectance slows down with increasing wavelength and reaches its highest value near 2200 nm. At this point, the slope of the curve approaches 0. In the 2200 to 2450 nm band range, the spectral reflectance slowly decreases, but the reflectance is still higher than the values in the 400 to 1000 nm

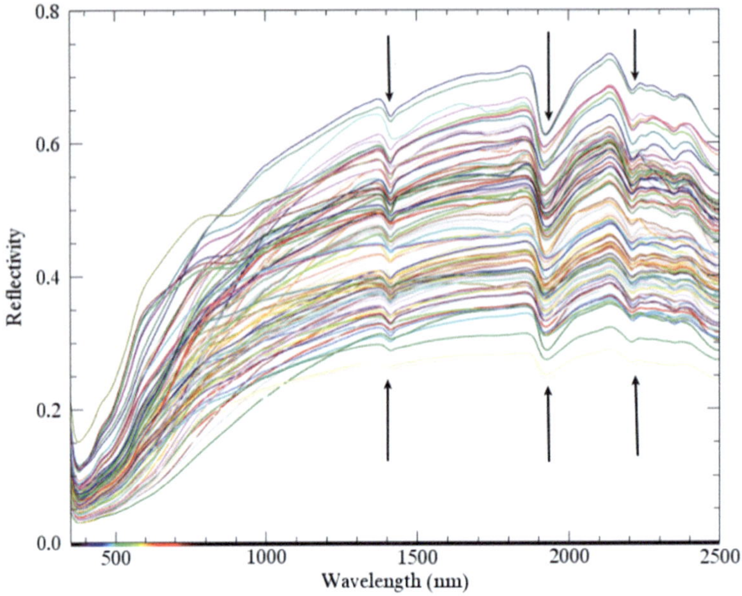

Fig. 4.6 Original spectral profile of TRSR soil after moving smoothing

band range. Also, three significant absorption valleys can be seen in the figure at the 1400 nm, 1900 nm, and 2200 nm bands, but they have different causes. 1400 nm and 1900 nm bands are mainly produced by the influence of small amounts of moisture and hydroxyl groups in the soil and air, while 2200 nm band is mainly produced by the influence of minerals and clay (Nan, 2017).

4.3.3.2 Soil Spectral Transformation Form Analysis

As the raw soil spectral data is noisy and in a single form, some mathematical transformations are used to expand the reflection and absorption characteristics between spectral bands, to uncover the hidden information between bands and to improve the sensitivity of soil nutrients to spectra. In this paper, seven transformations such as first-order differentiation (FD), second-order differentiation (SD), inverse logarithm (LR), de-envelope (CR), multiple scattering correction (MSC), and first-order differentiation based on de-envelope and multiple scattering correction (CRFD, MSCFD) are mainly used for subsequent modeling.

Figure 4.7a shows the spectral profile of the FD of the soil sample. As can be seen from the graph, the transformed spectral profile is highly variable, with reflectance values ranging roughly from − 0.0025 to 0.002, with most values fluctuating up and down around the value of 0. After the first-order differential transformation, the spectral reflectance values decrease and are concentrated around the value of 0, highlighting the information on the spectral absorption valley (Yang, 2014).

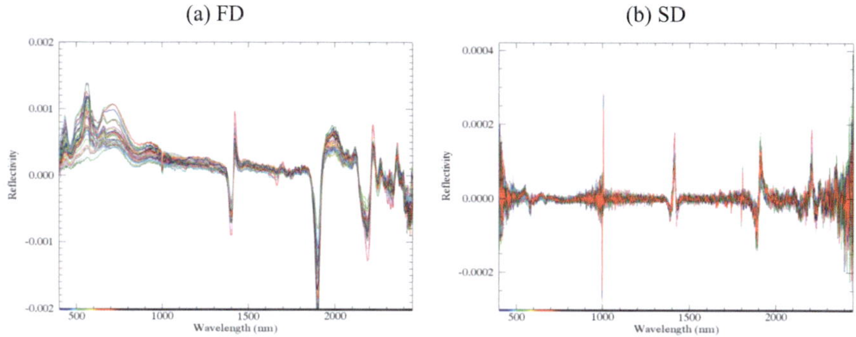

Fig. 4.7 Map of spectral transform FD, SD results for some soils in the TRSR area

Figure 4.7b shows the spectral profile of the SD of the soil sample. It can be seen that the spectral profile is highly variable after the transformation, with relatively sharp fluctuations up and down the curve. However, the differences in the spectral curves between the samples were significantly reduced and basically hovered around the reflectance value of 0 (Yang, 2014), with the spectral reflectance range roughly in the range of − 0.0003 to 0.0003. Compared to the first-order differential transform, the values are reduced by a factor of 10. The noise in the 400–500 nm band at the beginning of the curve and the 2400–2450 nm band at the end of the curve is obvious and may have some impact on the quality of the spectral data. The secondary differential transformed spectral curve highlights the band information of the original spectrum at the three peaks and valleys at 1000 nm, 1400 nm, 1900 nm, and 2200 nm (Yang, 2014).

Figure 4.8a shows the spectral profile of the LR of the soil sample. From it, it can be found that the transformed spectral reflectance ranges from 0.2 to 1.5. The transformation mainly highlights the spectral reflectance features in the range of 400–1000 nm, and expands the curve features in the 400–700 nm band. The curve features at the 1000–2450 nm band are weakened, including the absorption valley features at the 1400 nm, 1900 nm and 2200 nm bands, and the shape of the curve is flatter, with less noise influence (Yang, 2014).

Figure 4.8b shows the spectral profile of the CR of the soil sample. It can be found that the transformed spectral reflectance range is within the 0 to 1 interval. The transformation normalizes the spectral reflectance in the band range to a background, showing two main features. The spectral profile features in the 400–700 nm band range are enhanced and expanded, and the spectral profile features in the 1000–1800 nm band range are weakened. At the same time, the spectral profile that has undergone a de-envelope change has four significantly enhanced absorption valleys that help to extract valid information during the modeling process and improve the accuracy of the modeling. This is the commonly used form of spectral transformation.

Figure 4.9 shows the spectral profile of the MSC of the soil sample. It can be concluded that the transformed spectral curve rises steeply in the range of 400–1000 nm. It gradually slows down in the range of 1000–2200 nm, reaching a

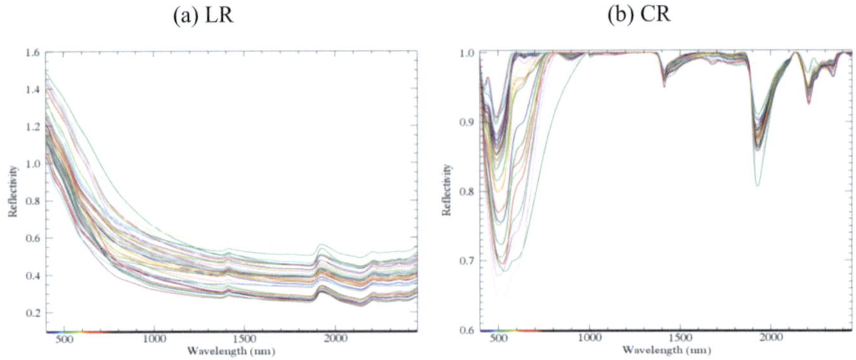

Fig. 4.8 Map of spectral transform LR, CR results for some soils in the TRSR area

maximum spectral reflectance at 2200 nm. The spectral curve decreases slowly in the 2200–2450 nm range (Yang, 2014). This form of transformation also enhances the spectral characteristics of the absorption valleys at 1400, 1900, and 2200 nm. When compared with the original spectral data, the spectral profile after the multiple scattering correction change is smoother and the spectral reflectance values are more concentrated. The spectral noise is greatly reduced (Yang, 2014).

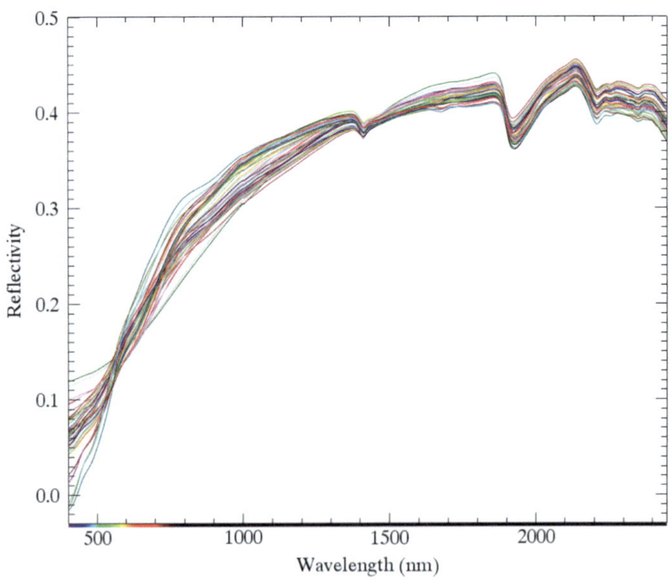

Fig. 4.9 Map of spectral transform MSC results for some soils in the TRSR area

4.3.3.3 Correlation Analysis Between Soil Nutrient Content and Spectral Reflectance

In order to be able to characterize the relationship between soil nutrient content and spectral reflectance, eight spectral variables were correlated with soil nutrient content. The correlation coefficient -r was calculated for each spectral variable and soil nutrient content and the correlation results were analyzed. The correlation analysis was completed in SPSS software and the correlation coefficient matrix is shown in Table 4.11. The results showed that there was a strong correlation between the soil nutrients. SOC, TN, TP, and TK were all significantly correlated at the 0.01 level, with the correlation coefficient between SOC and TN being as high as 0.687. This indicates that there is a dependency relationship between these soil nutrients.

Figure 4.10 show the correlation between soil nutrient content and spectral reflectance. It is observed that there is a consistent pattern between SOC, TN, TP, and TK content and spectral reflectance. In this paper, SOC is used as an example for analysis. Some patterns can be drawn from Fig. 4.10a, where the correlation between spectral reflectance and SOC is significantly enhanced by various transformations. The correlation between R and SOC content is negative, with correlation coefficients ranging from − 0.6 to − 0.05. The overall correlation curve is relatively flat, with an increasing trend in the visible range and minor troughs around 1400, 1900, and 2200 nm. The R and LR are symmetrically distributed, with correlation coefficient values in the LR form ranging from 0.05 to 0.6. The correlation coefficients of the FD and SD transformed forms are positively and negatively correlated with the SOC, with large fluctuations in the correlation coefficients. The correlation coefficients of the FD form increase in the visible near-infrared range. The correlation coefficients of the SD form are more variable, with faster rises and falls in the curve, and more intense absorption and reflection peaks. Thus the more information is presented. The correlation between the CR-treated spectra and the SOC content is high in the broad band, with a significant increase in the near-infrared range. The MSC form-transformed spectra show an increased correlation compared to R, with an increased number of absorption valleys and reflection peaks. Both CRFD and MSCFD are combinations of multiple transform forms, with a slight increase in correlation compared to one transform form alone. This suggests that the composite

Table 4.11 Correlation coefficient matrix for soil nutrient content

	SOC	TN	TP	TK
SOC	1			
TN	0.687**	1		
TP	0.312**	0.297**	1	
TK	− 0.129*	− 0.675**	0.039	1

Note ** indicates significant correlation at the 0.01 level. * indicates significant correlation at the 0.05 level

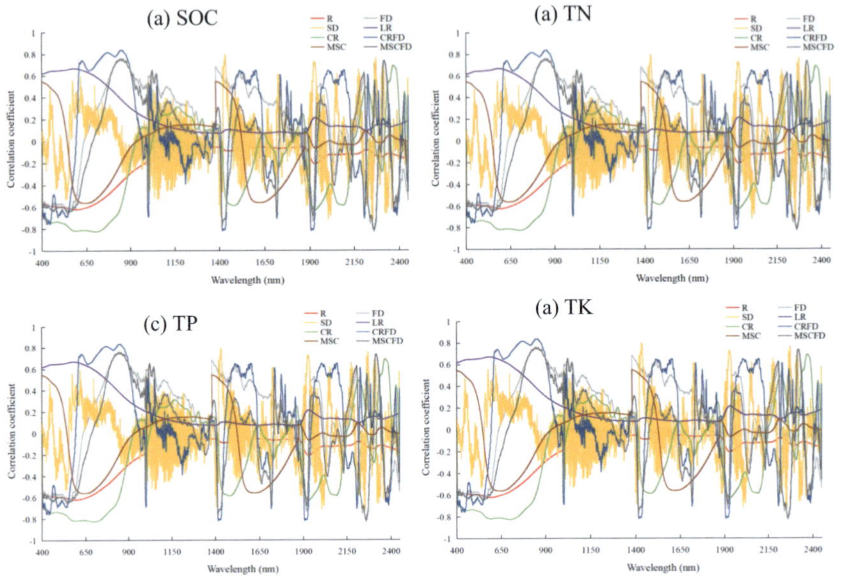

Fig. 4.10 Correlation curve between soil nutrient content and spectral data. *Note* (1) R is raw spectral reflectance. (2) FD is first order differential. (3) SD is second-order differential. (4) LR is inverse logarithm. (5) CR is de-envelope. (6) CRFD is de-envelope-first order differential. (7) MSC is multiple scattering correction. (8) MSCFD is multiple scattering correction-first order differential

mathematical transform has a better advantage in uncovering hidden information in the spectrum.

4.3.3.4 Soil Nutrient Characteristic Band Selection

In order to reduce the amount of data computation and improve computational efficiency, this paper uses the characteristic band method for modeling. The selection of soil nutrient characteristic bands should follow the principles of rich information content, large differences in the spectral characteristics of the features, and low correlation within the bands. The correlation between spectral variables and soil nutrients was analyzed and the bands with high correlation that passed the significance test at the P = 0.01 level were selected. Also, to improve the accuracy of the model estimation, it was decided to select bands with correlation coefficients above a certain threshold (SOC: $|r| \geq 0.6$; TN: $r \geq 0.5$, $r \leq -0.6$; TP: $r \geq 0.35$, $r \leq -0.5$; TK: $r \geq 0.35$, $r \leq -0.4$; Flexible selection of thresholds for special situations.) and are located at the positive and negative wave positions of the correlation curve. Since the 1400 nm and 1900 nm bands are more affected by moisture, two bands in the range of 1315–1415 nm and 1820–1940 nm were removed from the 400–2450 nm band. The significant bands between soil organic carbon, total nitrogen, total phosphorus, and total potassium, and the various forms of data transformation are shown in Table 4.12.

Table 4.12 Extraction of information on the characteristic bands of the soil spectrum corresponding to the soil nutrient elements for each form of data transformation

Significant bands of the different data transformations (nm)	Soil nutrients			
	SOC	TN	TP	TK
R	526, 593, 652	401, 1314, 1642 2131, 2216, 2358	–	2134, 2399
FD	549, 569, 859, 1025, 1515, 1538, 2239, 2258, 2318, 2386	811, 328, 856, 871, 889 2256, 2313, 2326, 2248	819, 1003, 1493, 1512, 2148, 2178, 2353	2013, 2065, 2089 2120, 2156, 2349
SD	1434, 1725, 2055 2174, 2273, 2312	997, 1004, 1649, 1728, 2017, 2124, 2261, 2262, 2312, 2330	418, 896, 1425 2015, 2174, 2201	1209, 1666, 1797, 1978, 2126, 2139, 2147, 2261
LR	402, 578, 773, 1941	403, 2132, 2217, 2360	–	2134, 2242, 2400
CR	455, 591, 699, 1482 1941, 2088, 2367	404, 410, 420, 430 473, 706, 2292, 2382	768, 2197, 2346	2128, 2168, 2269, 2318
CRFD	409, 443, 540, 626, 765 853, 1428, 1715, 2205, 2385	409, 613, 748, 780, 856, 1715, 2086, 2104, 2205, 2313	831, 1754 2244, 2360, 2425	1752, 1799, 1811 2242, 2289, 2314
MSC	400, 640, 1416, 1619	605, 1176, 1205 1590, 2215, 2353	522, 1502	1060, 1279, 2133, 2409
MSCFD	466, 549, 847, 876, 1025, 1416, 1715, 2206, 2257, 2316	799, 856, 894 2256, 2310, 2447	819, 1003, 1022 2149, 2178, 2353	1796, 1819, 2065 2077, 2089, 2157

Note (1) R is raw spectral reflectance. (2) FD is first-order differential. (3) SD is second-order differential. (4) LR is inverse logarithm. (5) CR is de-envelope. (6) CRFD is de-envelope-first order differential. (7) MSC is multiple scattering correction. (8) MSCFD is multiple scattering correction-first order differential. (9) Significant bands are bands in which the spectral indicators pass the significance test at the 0.01 level

Table 4.13 Comparison of SOC data feature band modeling results

Data transformation form	Model	Modeling set		Validation set			Data transformation Form	Model	Modeling set		Validation set		
		R^2_M	$RMSE_M$	R^2_V	$RMSE_V$	RPD_V			R^2_M	$RMSE_M$	R^2_V	$RMSE_V$	RPD_V
R	MLR	0.4699	16.9100	0.3891	17.4996	1.2661	CR	MLR	0.7729	11.0687	0.7643	11.1604	1.9852
	PLSR	0.4698	16.9124	0.3916	17.4510	1.2696		PLSR	0.7692	11.1589	0.7757	10.9006	2.0325
	SVM	0.6184	14.9017	0.5686	14.8102	1.4960		SVM	0.8165	10.5586	0.7870	10.3627	2.1380
	RF	0.9163	7.3694	0.5968	14.2325	1.5567		RF	0.9549	5.3491	0.8032	10.1266	2.1878
FD	MLR	0.7755	11.0041	0.7842	10.3615	2.1382	CRFD	MLR	0.8426	9.2149	0.8218	9.9568	2.2252
	PLSR	0.7430	11.7743	0.7725	10.6496	2.0804		PLSR	0.8408	9.2670	0.8221	9.8714	2.2444
	SVM	0.8886	8.3325	0.8312	9.1098	2.4320		SVM	0.8879	8.4337	0.8588	8.3380	2.6572
	RF	0.9726	4.2188	0.8597	8.4033	2.6365		RF	0.9713	4.2393	0.8737	8.0251	2.7608
SD	MLR	0.7770	10.9688	0.7929	10.2039	2.1713	MSC	MLR	0.5679	15.2673	0.5783	14.4864	1.5294
	PLSR	0.7752	11.0123	0.7915	10.2200	2.1678		PLSR	0.5679	15.2673	0.5783	14.4864	1.5294
	SVM	0.8402	9.7059	0.7589	10.9075	2.0312		SVM	0.7691	11.9481	0.7511	11.4813	1.9297
	RF	0.9622	4.8159	0.8411	8.9346	2.4797		RF	0.9498	5.7977	0.7927	10.1617	2.1803
LR	MLR	0.6210	14.2992	0.5813	14.4229	1.5361	MSCFD	MLR	0.7746	11.0267	0.7853	10.5635	2.0974
	PLSR	0.6210	14.2992	0.5813	14.4229	1.5361		PLSR	0.7421	11.7961	0.7891	10.5733	2.0954
	SVM	0.6956	13.5724	0.6217	14.1885	1.5615		SVM	0.8977	8.1702	0.8615	8.2808	2.6755
	RF	0.9171	7.4847	0.5562	14.7759	1.4994		RF	0.9787	3.6845	0.8739	8.1368	2.7229

Note The meaning of R, FD, SD, LR, CR, CRFD, MSC and MSCFD are same as Table 4.12. MLR is a multiple linear regression model. PLSR is a partial least squares regression model. SVM is a support vector machine model. RF is a random forest model. R^2 is the coefficient of determination. RMSE is the root mean square error. RPD is the relative analysis error

4.4 Discussion

Four methods, namely multiple linear regression (MLR), partial least squares (PLS), support vector machine (SVM), and random forest model (RF), were used to construct hyperspectral inversion models of soil nutrient contents under different forms of data transformation. The accuracy of all inversion models was evaluated to find the best model for soil nutrient content prediction in the TRSR area. In order to reflect the effect of soil nutrient content inversion models more visually and effectively, this study used the measured and predicted values from the modeling set and validation set to draw a scatter plot of measured-predicted values. In order to derive an assessment of soil nutrient prediction results, the best combination of spectral transformation and inversion model was obtained.

4.4.1 Results of SOC Content Characteristic Waveform Inversion

According to Table 4.6, among the four models, the modeling accuracy of the RF model is higher than that of the SVM, PLSR and MLR models. The modeling accuracy of the SVM is higher than that of the PLSR and MLR models, and the modeling accuracy of the PLSR is slightly higher than that of the MLR model. Looking at the accuracy evaluation metrics (R^2_M, R^2_V, $RMSE_M$, $RMSE_V$) of the modeling set and the validation set, it can be found that the R^2_M of the modeling set is higher than the R^2_V of the validation set for most of the data transformations, and the $RMSE_M$ of the modeling set is smaller than the $RMSE_V$ of the validation set. However, there are models in the FD, SD, CR, MSC, MSCFD, and other transformation forms where the accuracy of the validation set is better than that of the modeling set.

Secondly, of all the data transformations (R, FD, SD, LR, CR, CRFD, MSC, and MSCFD), the R-MLR and R-PLSR models had RPD values of less than 1.4 and could not predict SOC content. The R-SVM, R-RF, LR transformed form combination models, CR-MLR, MSC-MLR, MSC-PLSR, and MSC-SVM did not reach RPD values above 2.0 and could only provide rough estimates of SOC content. In contrast, the RPD values of the other models all reached above 2.0, indicating that the models have excellent predictive power. The model inversion after CRFD transformation was more effective, with higher accuracy in the modeling set and validation set for all four inversion models, followed by the MSCFD data transformation form. Among all the combinations of spectral transformations and inversion models, MSCFD-RF and CRFD-RF were the best fits in the modeling and validation sets, while R-MLR and R-PLSR were the worst fits, which may be related to the high noise content in the samples and the low number of feature bands. Therefore, the MSCFD-RF and CRFD-RF models can invert the SOC content of the TRSR area more accurately.

It can also be concluded that after the spectral transformation of the original spectral R, both R^2_M and $RMSE_M$ for the modeling set and R^2_V and $RMSE_V$ for the validation set are significantly improved. Taking the MLSR model as an example, the modeling set accuracy and validation set accuracy of the original spectral R are lower than those of the modeling set accuracy and validation set accuracy after the other data transformations. Moreover, the combination of multiple forms of data transformation produces more obvious results. In summary, the different ways of data transformation and different inversion models will affect the inversion accuracy of the final model.

The closer the sample points are to the straight line, the closer the actual measured value of the sample is to the predicted value, and the better the prediction of the model is (Yu et al., 2016). From Figs. 4.11, 4.12, 4.13, 4.14, 4.15, 4.16, 4.17, 4.18, it can be seen that the overall dispersion of sample points is low. In general, among the eight data transformations and 32 inversion models, the RF model and SVM model predicted better than the MLR model and PLSR model, and the MSCFD and CRFD spectral transformation forms predicted better than the other transformation forms. Therefore, the close trend between the measured and predicted values of MSCFD-RF and CRFD-RF models is significantly better than other combined models, and is more suitable for hyperspectral inversion of SOC content in the TRSR area.

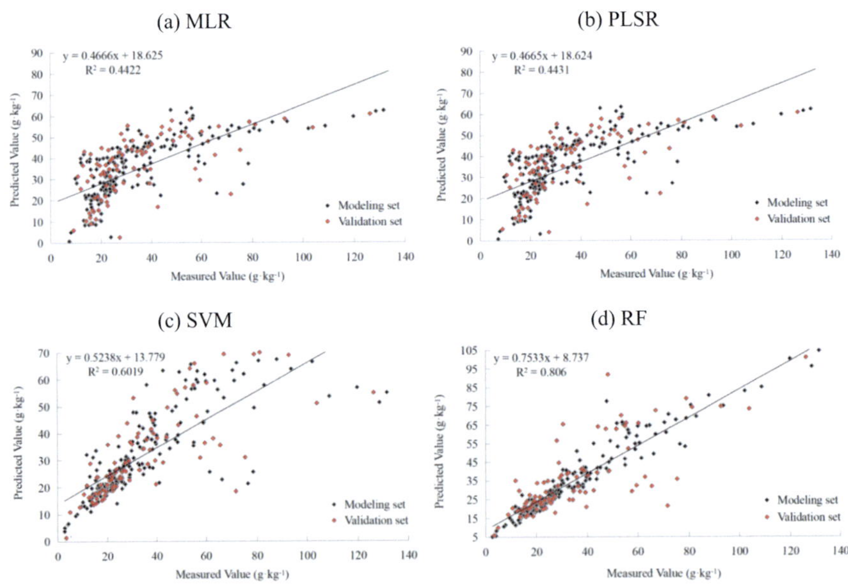

Fig. 4.11 Scatter plot of measured and predicted SOC content—R

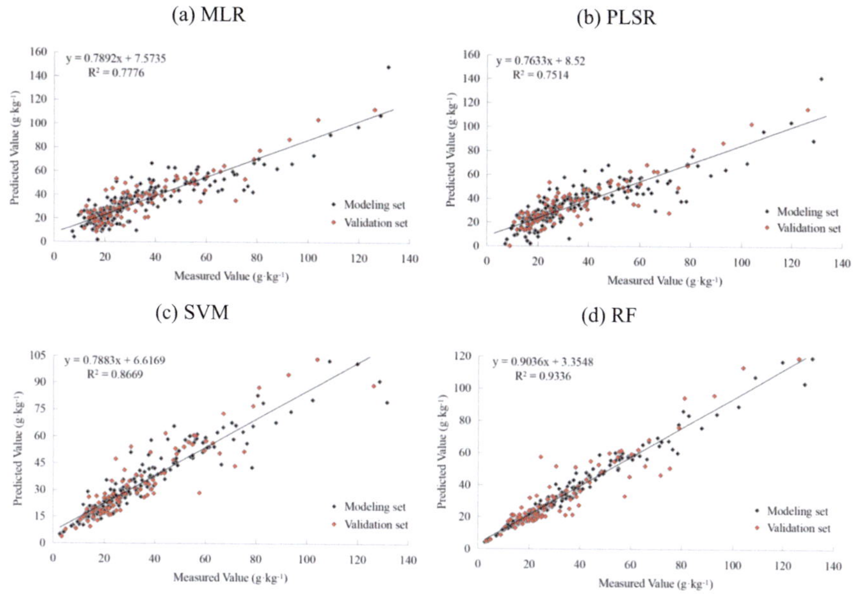

Fig. 4.12 Scatter plot of measured and predicted SOC content—FD

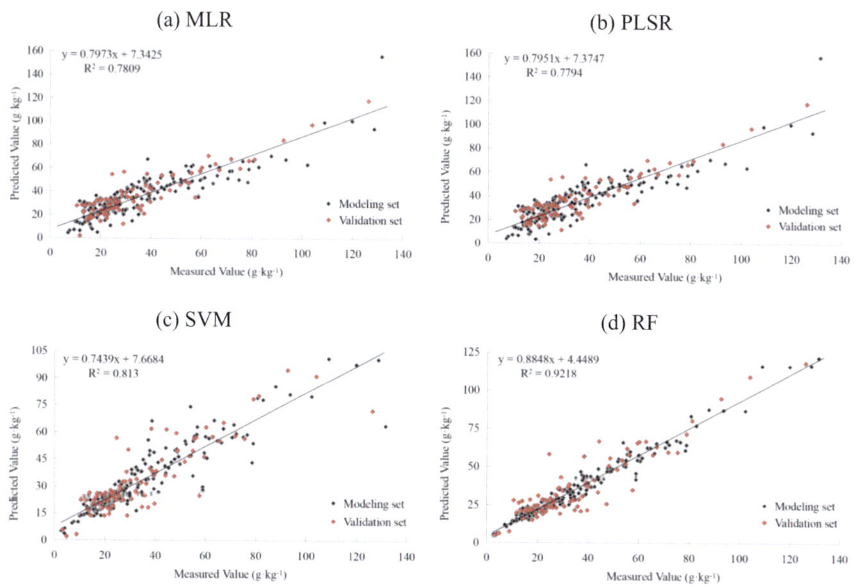

Fig. 4.13 Scatter plot of measured and predicted SOC content—SD

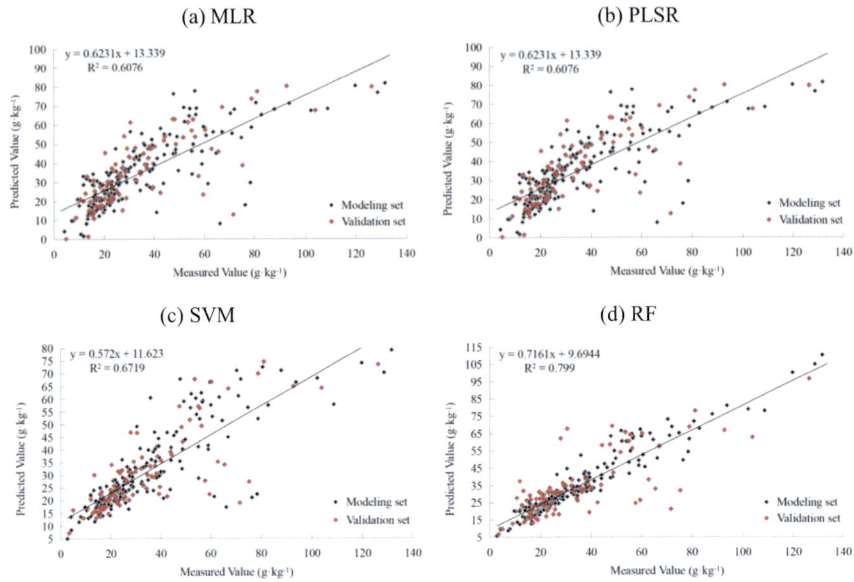

Fig. 4.14 Scatter plot of measured and predicted SOC content—LR

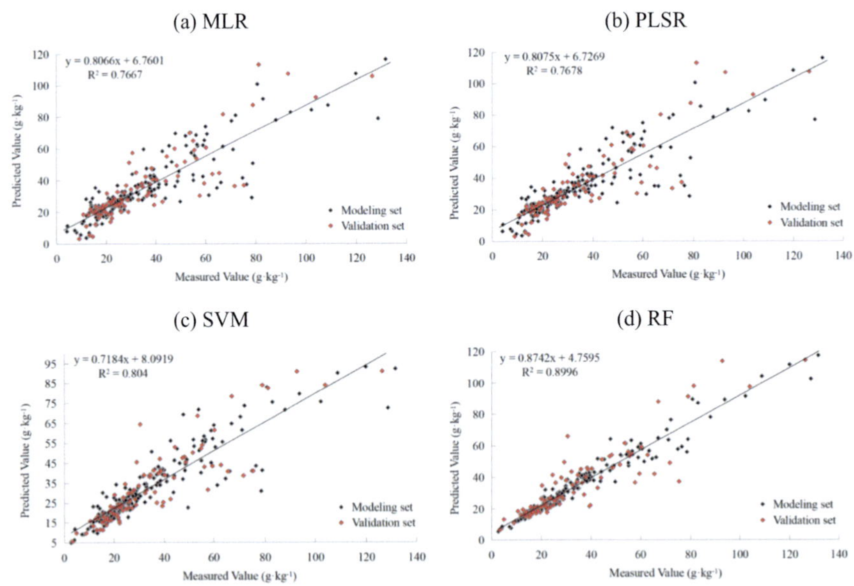

Fig. 4.15 Scatter plot of measured and predicted SOC content—CR

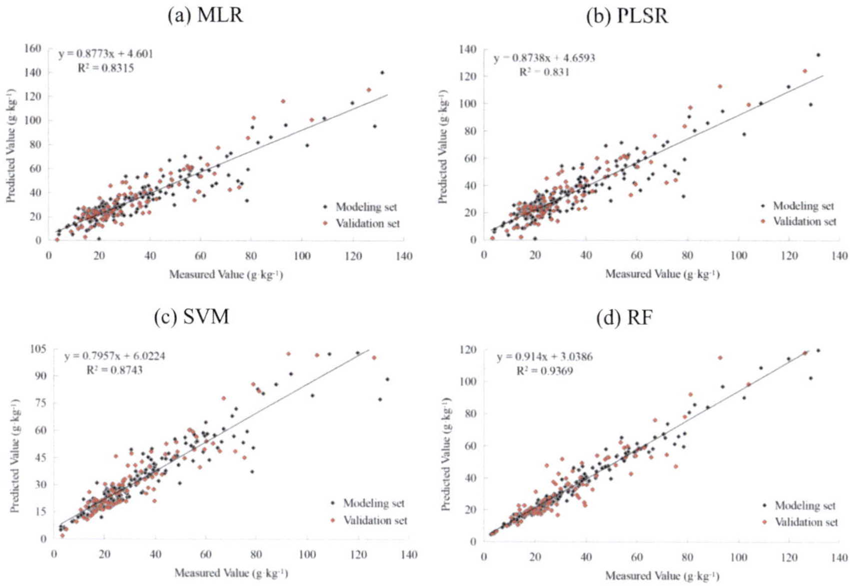

Fig. 4.16 Scatter plot of measured and predicted SOC content—CRFD

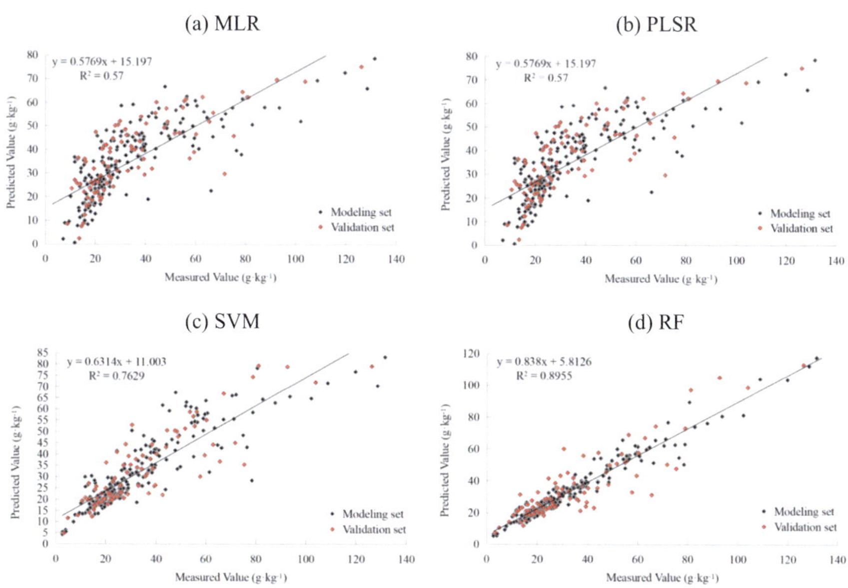

Fig. 4.17 Scatter plot of measured and predicted SOC content—MSC

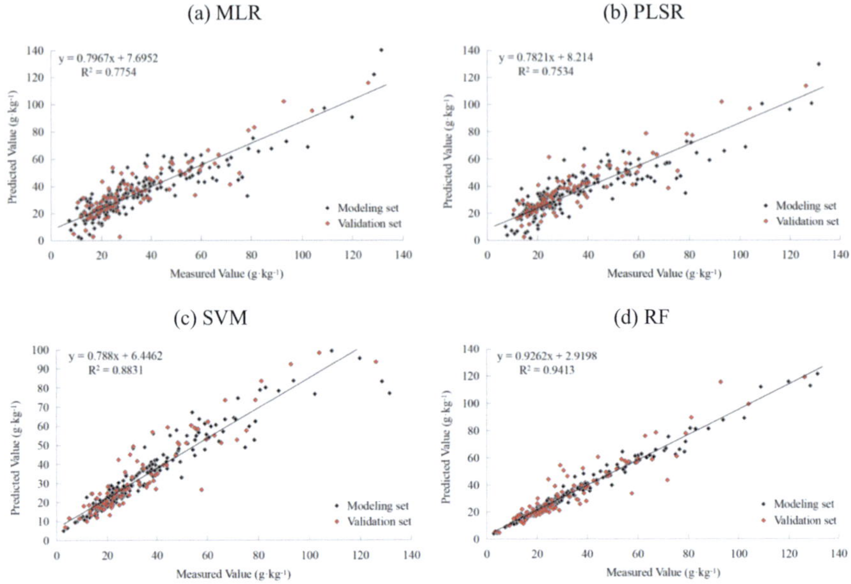

Fig. 4.18 Scatter plot of measured and predicted SOC content—MSCFD

4.4.2 Results of TN Content Characteristic Waveform Inversion

According to Table 4.14, among the four models, the RPD values of RF and SVM models are generally higher, indicating that the predictive ability of RF and SVM models is relatively good. The modeling set accuracy of RF models is higher than that of SVM, PLSR, and MLR models. The modeling set accuracy of SVM is higher than that of PLSR and MLR models in most cases, but the modeling set accuracy of SVM under the transformed form of R and LR set accuracy was lower in the transformed form of R and LR. The accuracy evaluation metrics (R^2_M, R^2_V, $RMSE_M$, $RMSE_V$) of the modeling set and validation set were observed, and it was found that the R^2_M of the modeling set was higher than the R^2_V of the validation set for most data transformations, and the $RMSE_M$ of the modeling set was smaller than the $RMSE_V$ of the validation set. But there were models in the FD, LR, CR, CRFD, MSC, MSCFD transformations where the accuracy of the validation set was higher than that of the modeling set Accuracy.

Secondly, of all the data transformations (R, FD, SD, LR, CR, CRFD, MSC, and MSCFD), the R-MLR, R-PLSR and R-SVM had RPD values less than 1.4 and could not predict total soil nitrogen content. The FD-RF, CRFD-SVM, CRFD-RF, MSC-SVM, MSC-RF, MSCFD- SVM, and MSCFD-RF models had RPD values greater than 0.2, indicating that these models had good predictive ability of TN content. The RPD values of the other models ranged from 1.4 to 2.0, indicating that these models

Table 4.14 Comparison of TN data feature band modeling results

Data transformation form	Model	Modeling set		Validation set			Data transformation Form	Model	Modeling set		Validation set		
		R^2_M	$RMSE_M$	R^2_V	$RMSE_V$	RPD_V			R^2_M	$RMSE_M$	R^2_V	$RMSE_V$	RPD_V
R	MLR	0.5291	1.4927	0.4676	1.5509	1.3630	CR	MLR	0.7016	1.1883	0.6328	1.3328	1.5860
	PLSR	0.5497	1.5076	0.4788	1.5324	1.3794		PLSR	0.6204	1.3402	0.6473	1.2844	1.6458
	SVM	0.5437	1.5308	0.5037	1.5215	1.3893		SVM	0.8104	0.9801	0.7245	1.1157	1.8947
	RF	0.9327	0.6820	0.6039	1.3487	1.5673		RF	0.9542	0.5372	0.6742	1.2233	1.7280
FD	MLR	0.6994	1.1926	0.6350	1.3019	1.6237	CRFD	MLR	0.6844	1.2221	0.7091	1.1602	1.8220
	PLSR	0.5907	1.3917	0.6338	1.2843	1.6459		PLSR	0.6806	1.2294	0.7011	1.1853	1.7833
	SVM	0.8868	0.8148	0.7559	1.0628	1.9889		SVM	0.9056	0.7213	0.8096	0.9761	2.1657
	RF	0.9643	0.4422	0.8031	0.9449	2.2372		RF	0.9727	0.7213	0.8152	0.9144	2.3118
SD	MLR	0.6817	1.2274	0.5832	1.3879	1.5231	MSC	MLR	0.5972	1.3807	0.6510	1.2620	1.6750
	PLSR	0.6227	1.3361	0.5943	1.3628	1.5511		PLSR	0.5711	1.4246	0.6527	1.2605	1.6770
	SVM	0.8379	0.9215	0.6244	1.3186	1.6031		SVM	0.7261	1.2080	0.8235	0.9592	2.2038
	RF	0.9577	0.5034	0.6493	1.2518	1.6887		RF	0.9618	0.4740	0.8518	0.8249	2.5627
LR	MLR	0.5566	1.4485	0.5047	1.4897	1.4189	MSCFD	MLR	0.6951	1.2011	0.6899	1.1981	1.7644
	PLSR	0.4511	1.6117	0.5041	1.4944	1.4145		PLSR	0.6295	1.3241	0.6696	1.2305	1.7179
	SVM	0.5733	1.4496	0.5424	1.4494	1.4584		SVM	0.8756	0.8229	0.8250	0.8934	2.3660
	RF	`0.9201	0.7542	0.5896	1.3789	1.5329		RF	0.9686	0.4227	0.8592	0.8034	2.6313

Note The meaning of all abbreviation are same as Table 4.13

could only provide rough estimates of the TN content. Among the eight spectral transformation forms, the inversion models after CRFD and MSCFD data transformation were more effective, and the modeling accuracy and validation accuracy of the four inversion models were relatively high. MSC and FD data transformation forms were the second most effective. Among all combinations of spectral transformation forms and inversion models, the best fit in the modeling set is CRFD-RF, and the best stability in the validation set is MSCFD-R. The better prediction results are MSCFD-RF ($RPD_V = 2.6313$) and MSC-RF ($RPD_V = 2.5627$). Therefore, the MSCFD-RF and MSC-RF models are more effective for inversion of TN content in the TRSR area.

In addition to this, for TN, the soil spectral profiles were data transformed to higher accuracy than the original spectra, both for R^2_M and $RMSE_M$ in the modeling set and R^2_V and $RMSE_V$ in the validation set, in line with the pattern for SOC. Except for LR, the form of data transformation does not conform to the above pattern, which is mainly due to the low number of characteristic bands.

From Figs. 4.19, 4.20, 4.21, 4.22, 4.23, 4.24, 4.25 and 4.26, the sample points are generally concentrated. 8 data transformations, out of 32 inversion models, the RF model outperforms the SVM, MLR, and PLSR models in terms of prediction. The three spectral transformations, MSCFD, CRFD, and MSC, outperformed the other transformations. In summary, the closeness of the measured and predicted values of the MSCFD-RF and MSC-RF models is significantly better than the other combined models, and their inversion effects can be a good estimation of the TN content in the TRSR area.

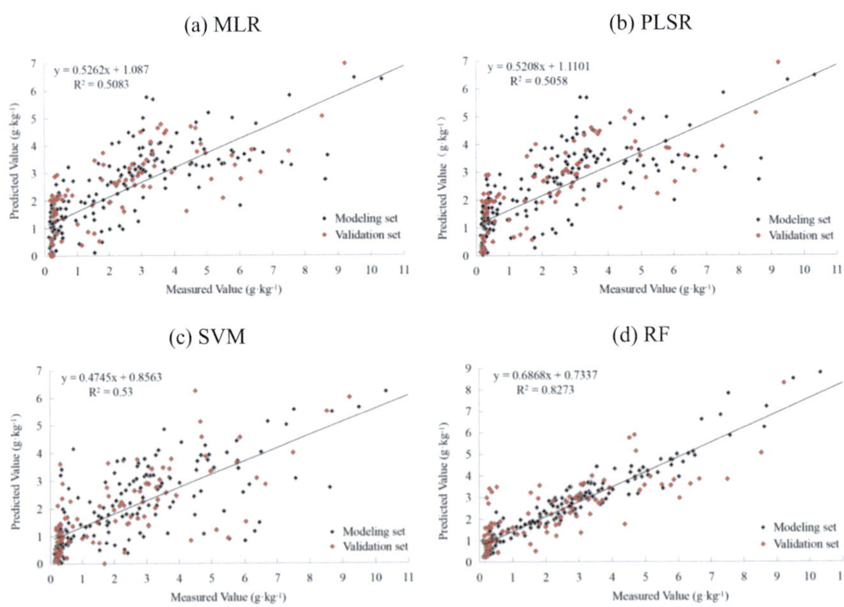

Fig. 4.19 Scatter plot of measured and predicted TN content—R

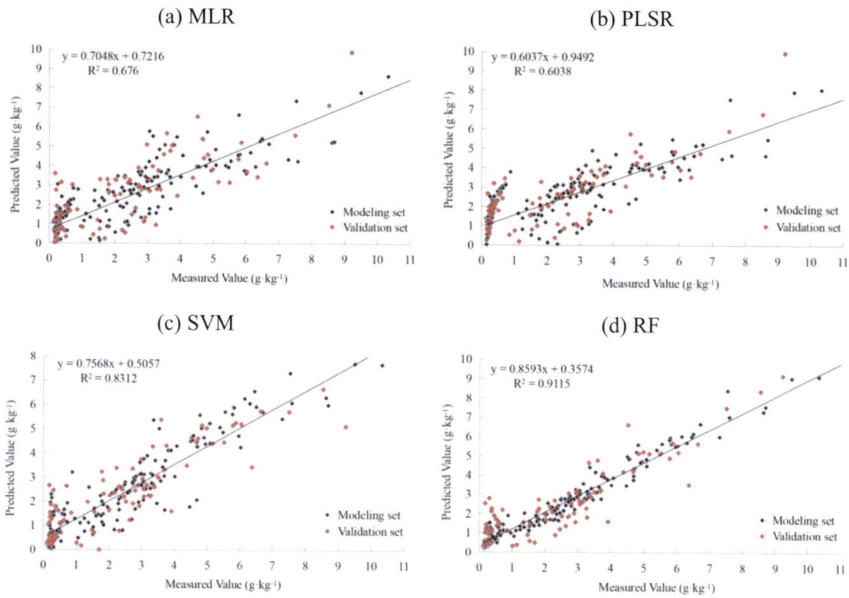

Fig. 4.20 Scatter plot of measured and predicted TN content—FD

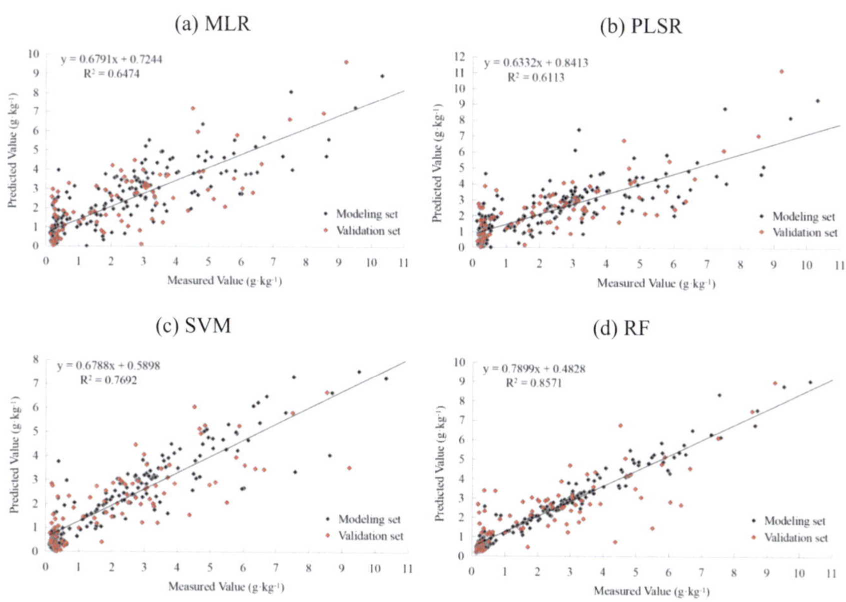

Fig. 4.21 Scatter plot of measured and predicted TN content—SD

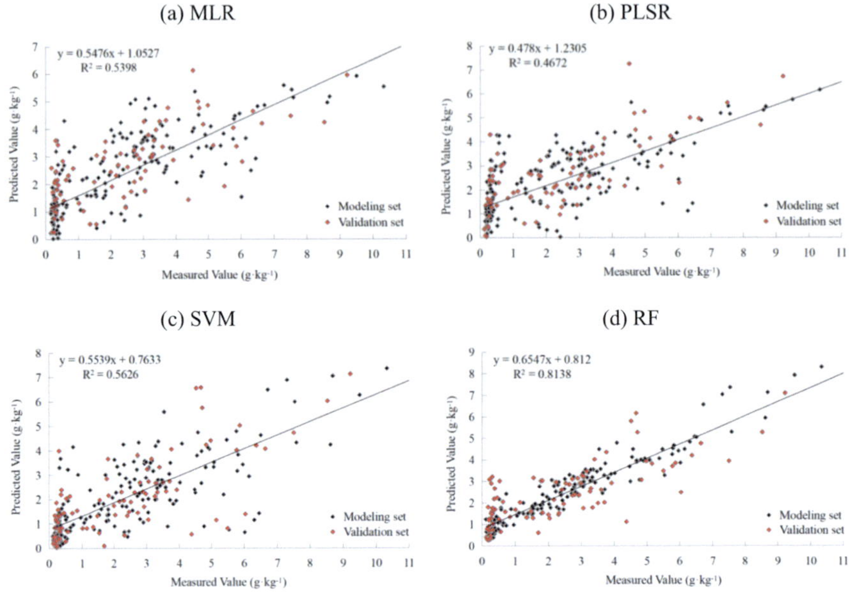

Fig. 4.22 Scatter plot of measured and predicted TN content—LR

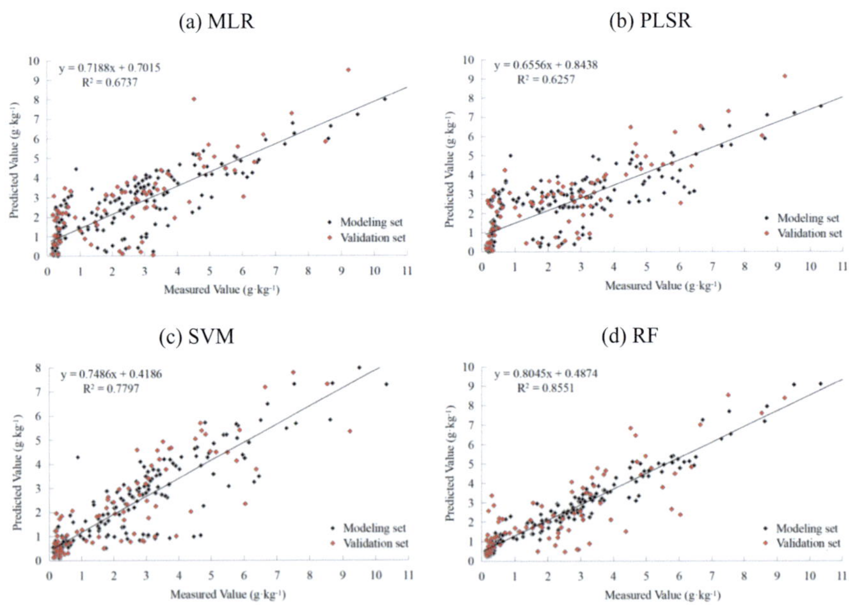

Fig. 4.23 Scatter plot of measured and predicted TN content—CR

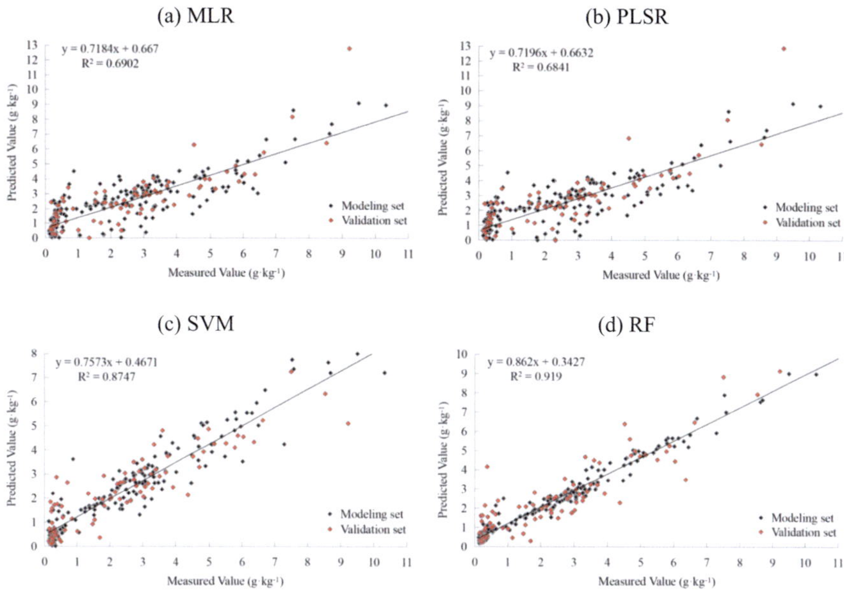

Fig. 4.24 Scatter plot of measured and predicted TN content—CRFD

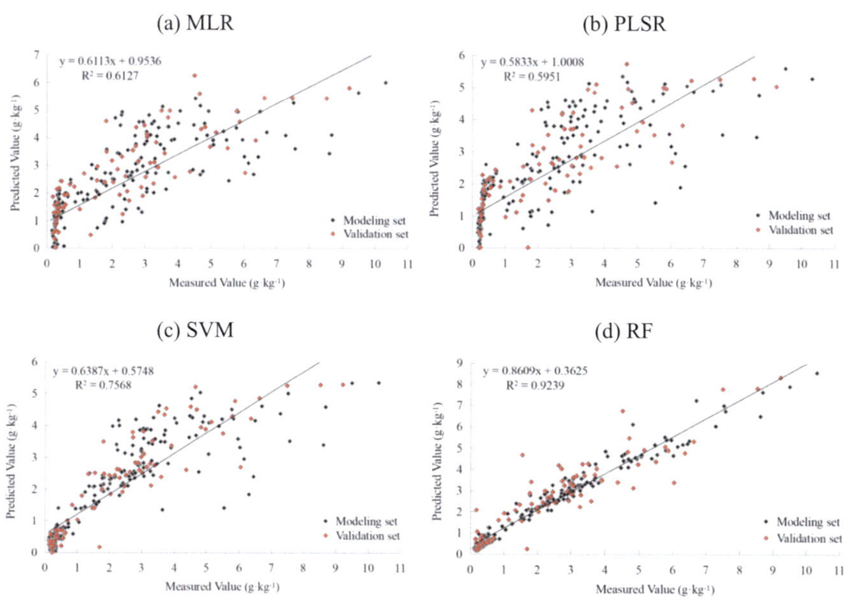

Fig. 4.25 Scatter plot of measured and predicted TN content—MSC

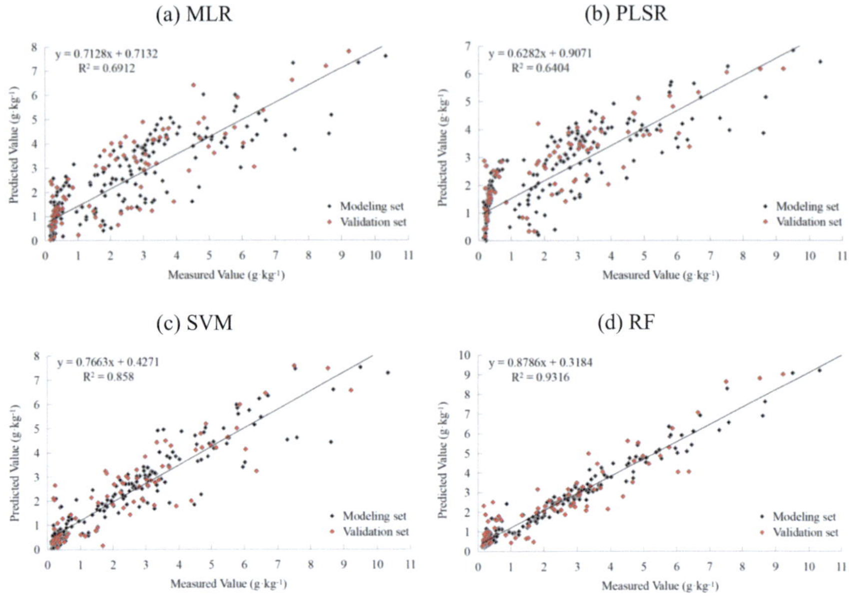

Fig. 4.26 Scatter plot of measured and predicted TN content—MSCFD

4.4.3 *Results of TP Content Characteristic Waveform Inversion*

According to Table 4.15, it can be seen that among the four models, the modeling set accuracy of the RF model is higher than that of the SVM, PLSR, and MLR models, and the modeling set accuracy of the SVM is higher than that of the PLSR and MLR models. Looking at the accuracy evaluation metrics (R^2_M, R^2_V, $RMSE_M$, $RMSE_V$) of the modeling set and the validation set, it is found that the modeling set R^2_M of all data transformations is higher than the R^2_V of the validation set, and the $RMSE_M$ of the modeling set is smaller than the $RMSE_V$ of the validation set (Table 4.15).

In addition, the R and LR data transformation forms had no eigenbands that passed the $p = 0.01$ significance test and therefore model inversion could not be performed. Of the remaining data transformations (FD, SD, CR, CRFD, MSC, and MSCFD), the modeling set accuracy and validation set accuracy of the four inversion models in the FD and MSCFD data transformation forms were higher among all models. However, the RPD values for all 24 models were less than 1.4, indicating that these models were unable to predict TP content in the TRSR area. This is mainly due to the low values of TN content.

From Figs. 4.27, 4.28, 4.29, 4.30, 4.31 and 4.32, it can be seen that of the six data transformations and 24 inversion models, the modeling results of all models are less than satisfactory due to the overall dispersion of sample points. The poor application significance of the results is largely due to the low level of TP in the TRSR region.

Table 4.15 Comparison of TP data feature band modelling results

Data transformation Form	Model	Modeling set		Validation set			Data transformation Form	Model	Modeling set		Validation set		
		R^2_M	$RMSE_M$	R^2_V	$RMSE_V$	RPD_V			R^2_M	$RMSE_M$	R^2_V	$RMSE_V$	RPD_V
R	–	–	–	–	–	–	CR	MLR	0.1472	0.1945	0.0514	0.1977	1.0093
	–	–	–	–	–	–		PLSR	0.1424	0.1950	0.0467	0.1986	1.0048
	–	–	–	–	–	–		SVM	0.2254	0.1886	0.0333	0.2018	0.9889
	–	–	–	–	–	–		RF	0.8030	0.1130	0.0455	0.2015	0.9904
FD	MLR	0.2599	0.1812	0.1795	0.1829	1.0914	CRFD	MLR	0.1982	0.1886	0.0806	0.1950	1.0238
	PLSR	0.2569	0.1815	0.1818	0.1824	1.0943		PLSR	0.1252	0.1970	0.0861	0.1910	1.0448
	SVM	0.3744	0.1708	0.1147	0.1906	1.0470		SVM	0.3395	0.1743	0.0667	0.1982	1.0071
	RF	0.9006	0.0868	0.1197	0.1898	1.0515		RF	0.9016	0.0952	0.0803	0.1934	1.0320
SD	MLR	0.2587	0.1813	0.0622	0.2015	0.9908	MSC	MLR	0.1381	0.1955	0.0432	0.1980	1.0081
	PLSR	0.2062	0.1876	0.0628	0.1969	1.0136		PLSR	0.0741	0.2026	0.0640	0.1933	1.0327
	SVM	0.4109	0.1643	0.1053	0.1914	1.0428		SVM	0.3646	0.1695	0.0352	0.2074	0.9625
	RF	0.8859	0.0887	0.1409	0.1896	1.0526		RF	0.8656	0.0903	0.1143	0.2002	0.9970
LR	–	–	–	–	–	–	MSCFD	MLR	0.2404	0.1835	0.1867	0.1814	1.1003
	–	–	–	–	–	–		PLSR	0.2090	0.1873	0.1561	0.1843	1.0831
	–	–	–	–	–	–		SVM	0.4263	0.1643	0.1190	0.1913	1.0434
	–	–	–	–	–	–		RF	0.9015	0.0847	0.1235	0.1912	1.0441

Note (1) R is raw spectral reflectance. (2) FD is first order differential. (3) SD is second order differential. (4) LR is inverse logarithm. (5) CR is de-envelope. CRFD is de-envelope-first order differential. (7) MSC is multiple scattering correction. (8)MSCFD is multiple scattering correction-first order differential. (9) MLR is a multiple linear regression model. (10) PLSR is a partial least squares regression model. (11) SVM is a support vector machine model. (12) RF is a random forest model. (13) R^2 is the coefficient of determination. (14) RMSE is the root mean square error. (15) RPD is the relative analysis error

Table 4.16 Comparison of TK data feature band modelling results

Data transformation form	Model	Modeling set		Validation set			Data transformation form	Model	Modeling set		Validation set		
		R^2_M	$RMSE_M$	R^2_V	$RMSE_V$	RPD_V			R^2_M	$RMSE_M$	R^2_V	$RMSE_V$	RPD_V
R	MLR	0.1987	6.5927	0.1792	6.6635	1.0963	CR	MLR	0.6553	4.3237	0.6044	4.6002	1.5880
	PLSR	0.1842	6.6520	0.1717	6.6853	1.0927		PLSR	0.5337	5.0289	0.6333	4.4855	1.6286
	SVM	0.1933	6.9010	0.1457	7.1290	1.0247		SVM	0.7980	3.3343	0.8188	3.1177	2.3431
	RF	0.8312	3.4139	0.1729	6.8297	1.0696		RF	0.9622	1.5326	0.8285	3.0678	2.3812
FD	MLR	0.5618	4.8751	0.5417	4.9973	1.4618	CRFD	MLR	0.5772	4.7886	0.5231	5.0812	1.4377
	PLSR	0.5559	4.9085	0.5643	4.8753	1.4984		PLSR	0.5629	4.8690	0.5274	5.0445	1.4481
	SVM	0.6641	4.3767	0.7062	4.0111	1.8212		SVM	0.8647	2.7138	0.7737	3.5041	2.0847
	RF	0.9296	2.1319	0.6858	4.1328	1.7676		RF	0.9589	1.6033	0.7825	3.4656	2.1078
SD	MLR	0.6852	4.1325	0.5809	4.8328	1.5116	MSC	MLR	0.5315	5.0412	0.5198	5.0689	1.4412
	PLSR	0.6776	4.1818	0.5991	4.6975	1.5551		PLSR	0.5315	5.0413	0.5190	5.0730	1.4400
	SVM	0.8745	2.6225	0.7760	3.5032	2.0852		SVM	0.7603	3.6516	0.7167	4.0341	1.8108
	RF	0.9606	1.5881	0.7726	3.4981	2.0883		RF	0.9754	1.2370	0.8865	2.4899	2.9339
LR	MLR	0.4574	5.4252	0.4945	5.2068	1.4030	MSCFD	MLR	0.5626	4.8711	0.6083	4.5883	1.5921
	PLSR	0.4574	5.4252	0.4945	5.2068	1.4030		PLSR	0.5595	4.8879	0.6028	4.6158	1.5826
	SVM	0.3199	6.1050	0.2921	6.2309	1.1724		SVM	0.8135	3.1852	0.8153	3.1437	2.3237
	RF	0.8772	2.9854	0.2799	6.2609	1.1668		RF	0.9564	1.6143	0.7647	3.5531	2.0559

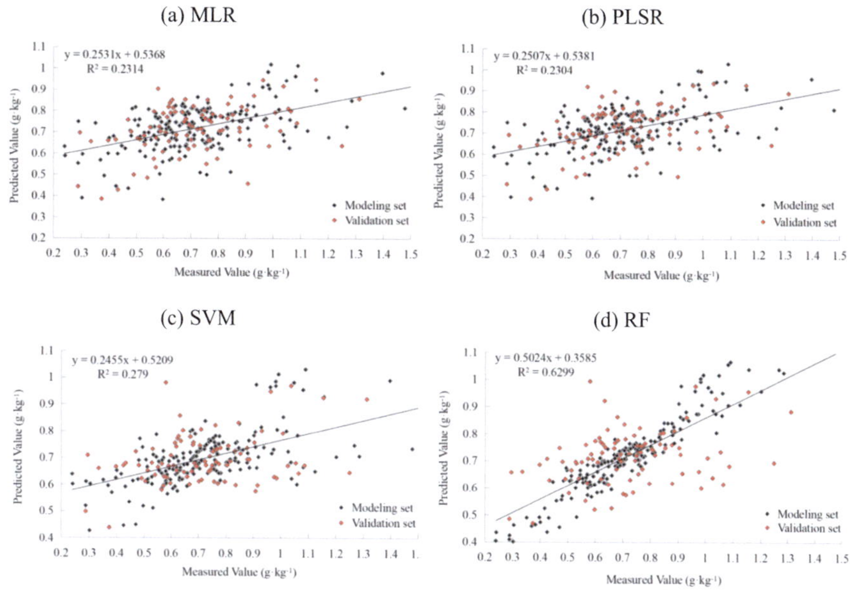

Fig. 4.27 Scatter plot of measured and predicted TP content—FD

However, the reliability of the various finger data transformation forms for exploring spectral information can be effectively evaluated by comparing the accuracy of all the spectra constructed models.

4.4.4 Results of TK Content Characteristic Waveform Inversion

According to Table 4.16, the modeling set accuracy of the RF model is higher than that of the SVM, PLSR and MLR models for all four models. The modeling set accuracy of the SVM is generally higher than that of the PLSR and MLR models, except for the LR data transformation form. Looking at the accuracy evaluation metrics ($R^2{}_M$, $R^2{}_V$, $RMSE_M$, $RMSE_V$) of the modeling set and validation set, it was found that the accuracy of the modeling set under most data transformations was higher than that of the validation set. However, there are models in the FD, LR, CR and MSCFD transformations where the validation set accuracy is higher than the modeling set accuracy.

Second, of all the data transformations (R, FD, SD, LR, CR, CRFD, MSC, and MSCFD), only the SD-SVM, SD-RF, CR-SVM, CR-RF, CRFD-SVM, CRFD-RF, MSC-RF, MSCFD-SVM, and MSCFD-RF models had RPD values of 2.0 or more, indicating that these models can predict the TK content of the TRSR area extremely well. The RPD values of R-MLR, R-PLSR, R-SVM, R-RF, and LR-SVM were less

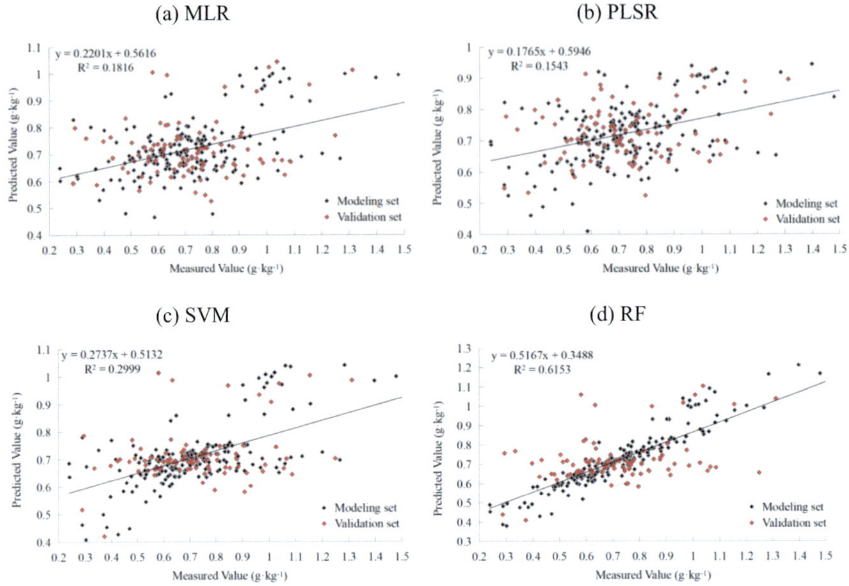

Fig. 4.28 Scatter plot of measured and predicted TP content—SD

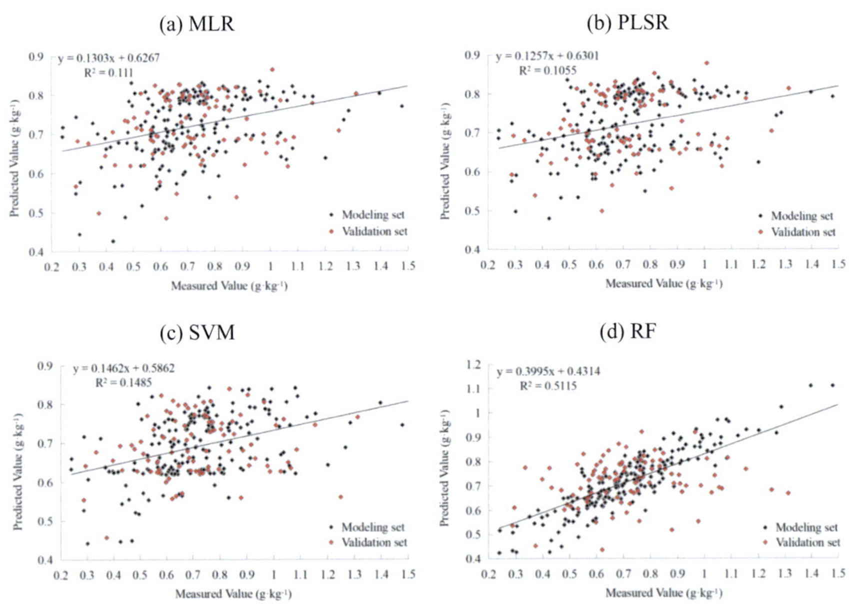

Fig. 4.29 Scatter plot of measured and predicted TP content—CR

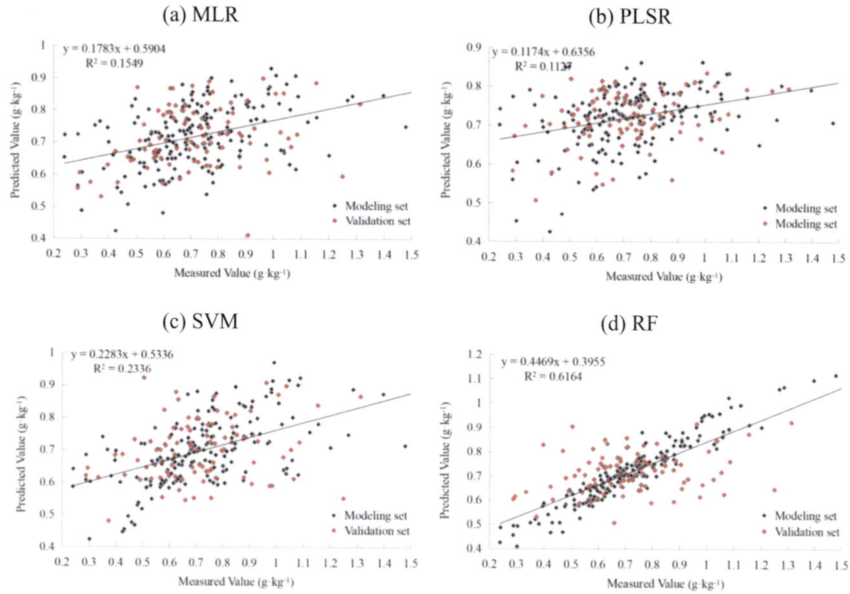

Fig. 4.30 Scatter plot of measured and predicted TP content—CRFD

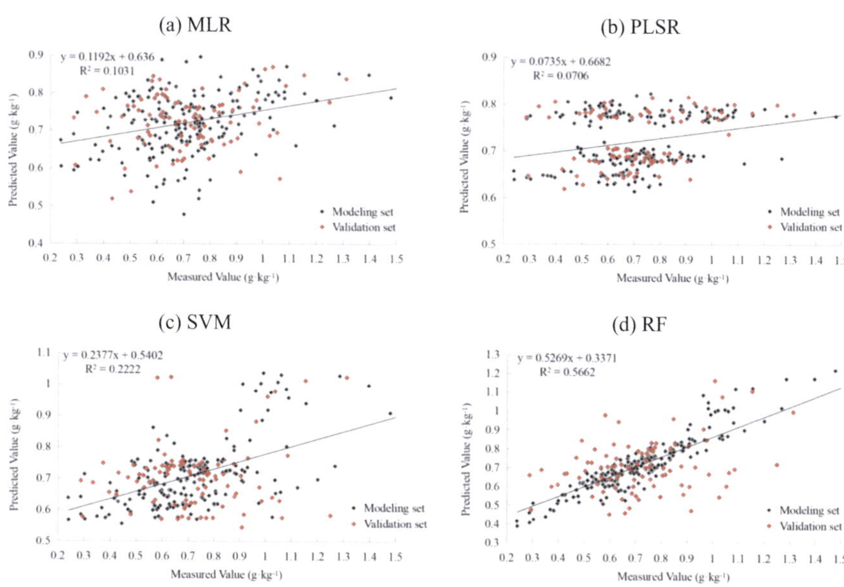

Fig. 4.31 Scatter plot of measured and predicted TP content—MSC

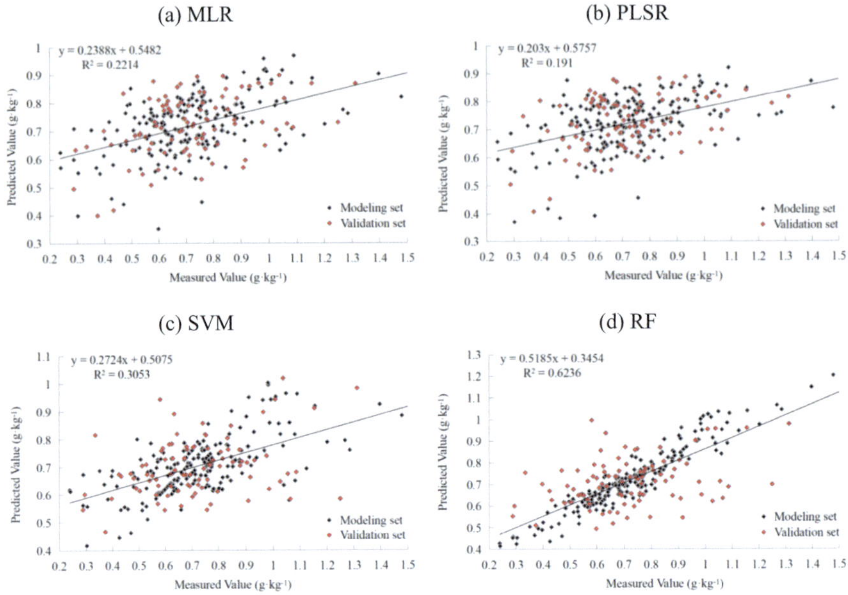

Fig. 4.32 Scatter plot of measured and predicted TP content—MSCFD

than 1.4, indicating that these models were not able to predict TK content. The other models had RPD values between 1.4 and 2.0, and could only roughly estimate TK content. the inversion effect models under the CR data transformation were better, and the modeling set accuracy and validation set accuracy were higher for all four inversion models. The MSCFD data-transformed form was the next best. Among all combinations of spectral transformations and inversion models, the best fit was MSC-RF in both the modeling and validation sets, and the worst fit was R-SVM. The models with better predictive power were MSC-RF ($RPD_V = 2.9339$), CR-RF ($RPD_V = 2.3812$), CR-RF ($RPD_V = 2.3431$), MSCFD-SVM $RPD_V = 2.3237$). Therefore, the MSC-RF and CR-RF models are more effective for inversion of TK content in the TRSR area.

At the same time, the accuracy of the soil spectral profile was slightly improved by the data transformation, both for R^2_M and $RMSE_M$ in the modeling set, and R^2_V and $RMSE_V$ in the validation set. In the SVM model, for example, the accuracy of the modeling set and the accuracy of the validation set of *R* are both lower than those of the modeling set and the validation set after other data transformations. Moreover, the combination of multiple data transformations produced more significant results.

From Figs. 4.33, 4.34, 4.35, 4.36, 4.37, 4.38, 4.39 and 4.40, it can be seen that the overall dispersion of sample points is small. 8 data transformations, among 32 inversion models, the prediction effect of RF model is better than that of SVM model, MLR model and PLSR model, and the prediction effect of CR and MSCFD spectral transformation forms is better than other transformations. In summary, the MSC-RF

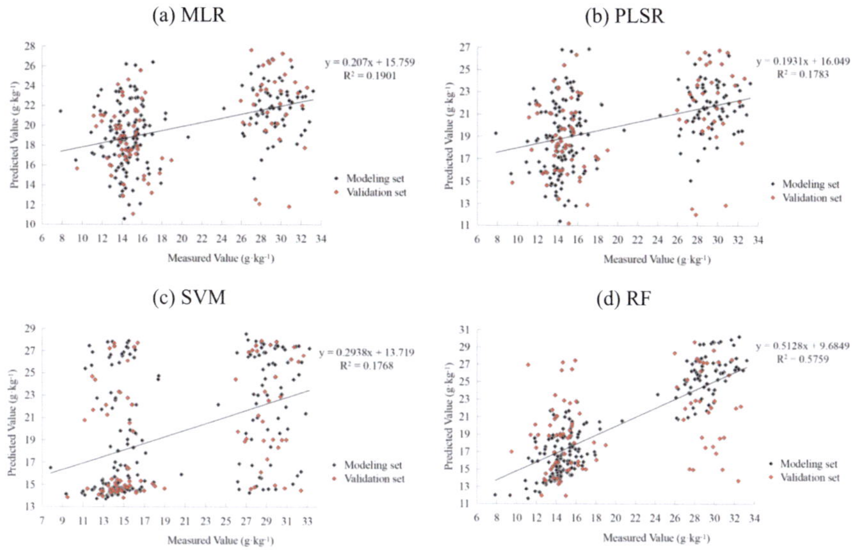

Fig. 4.33 Scatter plot of measured and predicted TK content—R

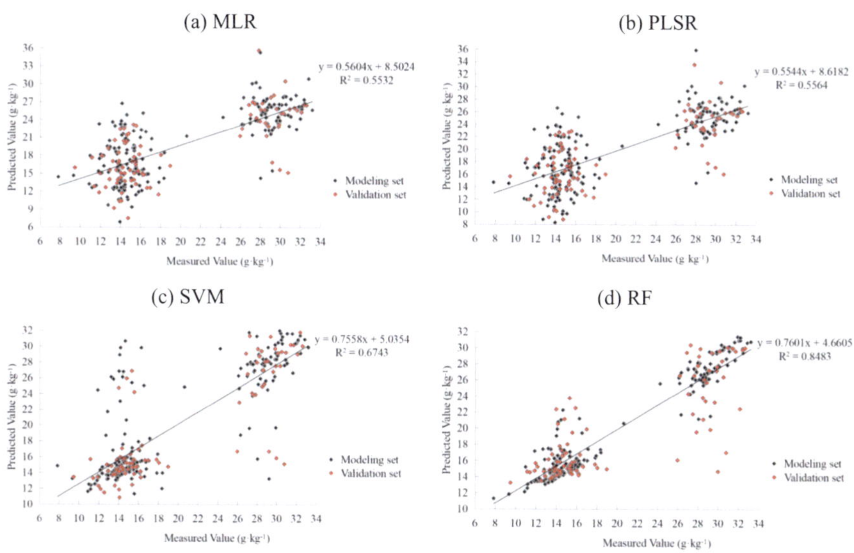

Fig. 4.34 Scatter plot of measured and predicted TK content—FD

and CR-RF models were used for hyperspectral inversion models of TK content in the TRSR area, and the results obtained were more satisfactory.

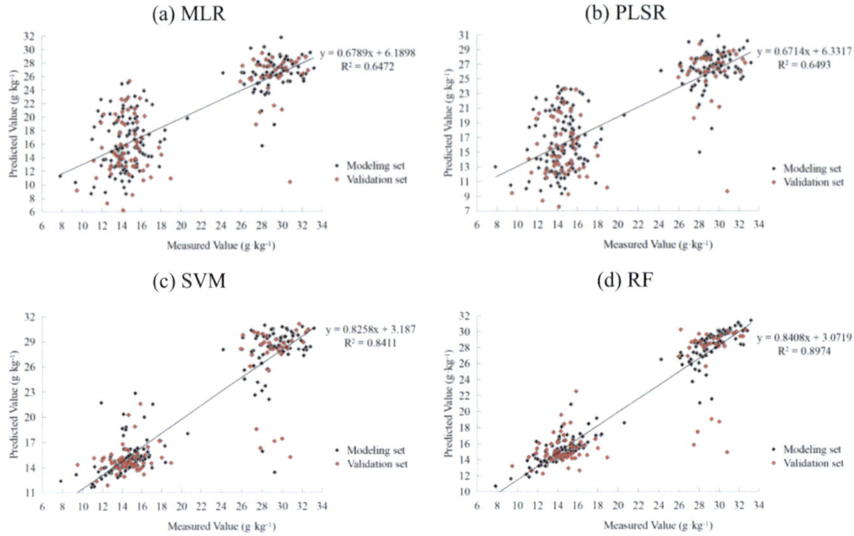

Fig. 4.35 Scatter plot of measured and predicted TK content—SD

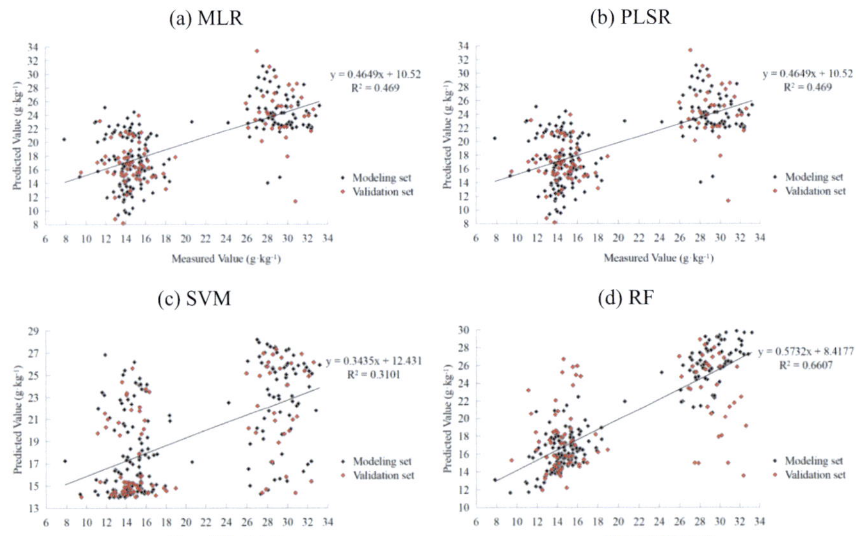

Fig. 4.36 Scatter plot of measured and predicted TK content—LR

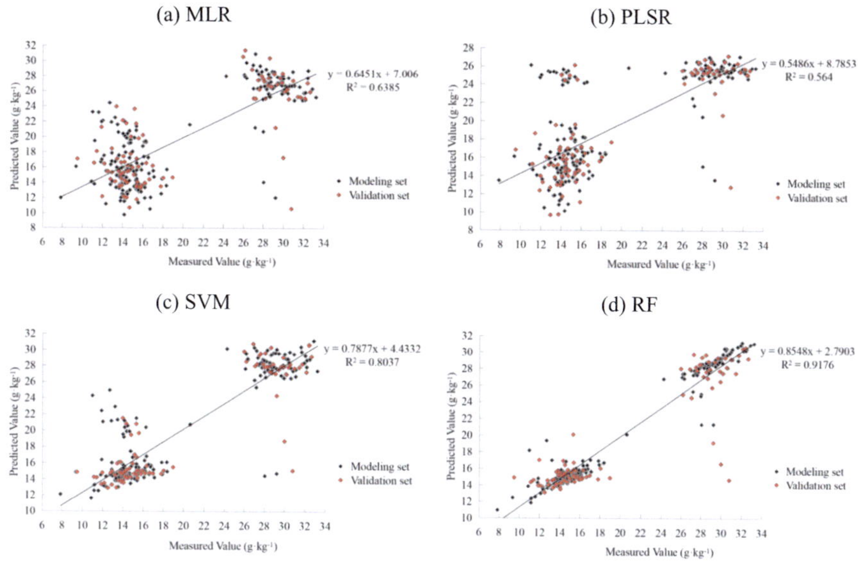

Fig. 4.37 Scatter plot of measured and predicted TK content—CR

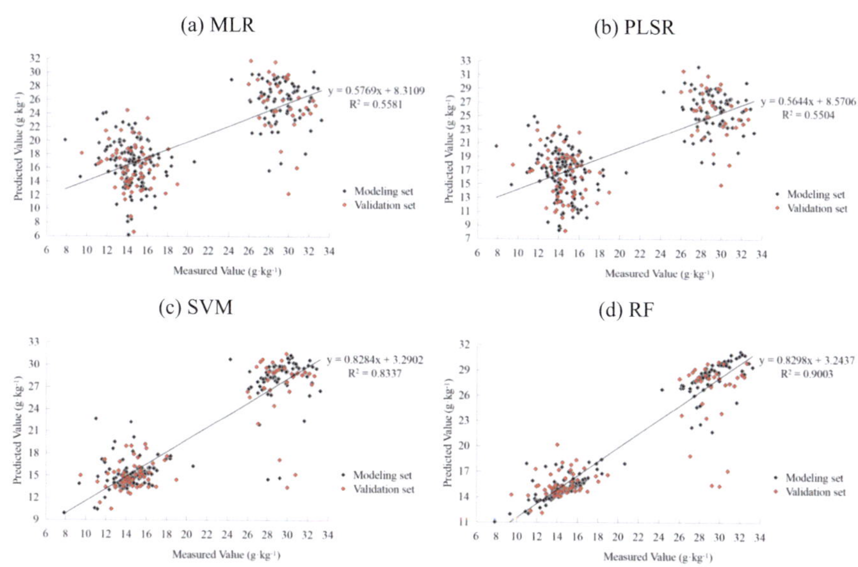

Fig. 4.38 Scatter plot of measured and predicted TK content—CRFD

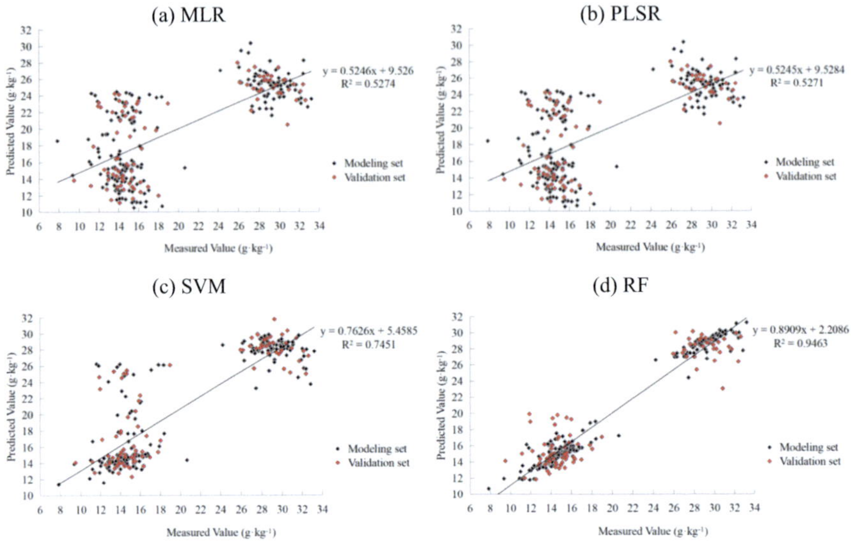

Fig. 4.39 Scatter plot of measured and predicted TK content—MSC

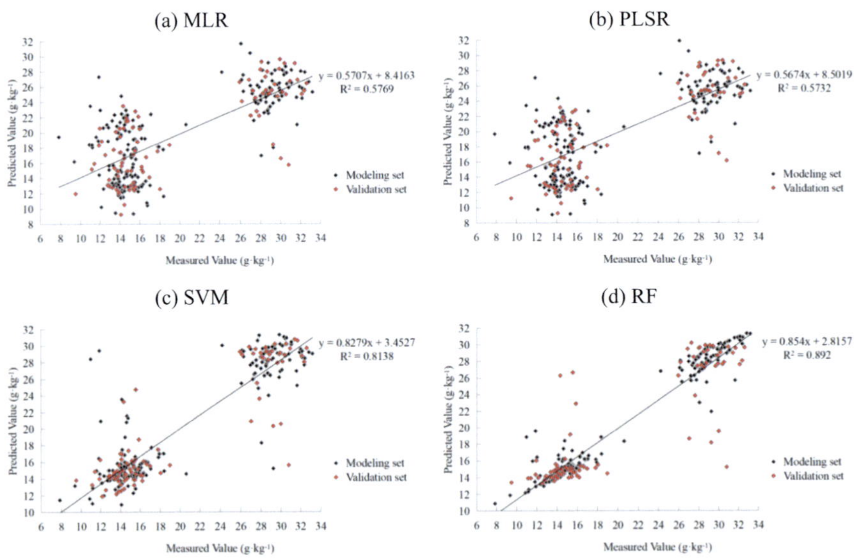

Fig. 4.40 Scatter plot of measured and predicted TK content—MSCFD

4.5 Conclusions

We used hyperspectral remote sensing data and applied MSLR, PLSR, SVM, and RF models to invert the contents of SOC, TN, TP, and TK in soils of the TRSR area. The following are the main conclusions drawn from this study. (1) The average contents of SOC,, TN, TP, and TK vary relatively widely in the TRSR area. The soil nutrient content was higher in vegetation types with high vegetation cover, more moisture and higher vegetation dead leaf volume, but was also influenced by the degree of vegetation dispersion, and the degree of rodent infestation. (2) There is consistency in the pattern, trend and location of the main peaks and troughs in the spectral reflectance curves for all soil samples. The spectral reflectance values for all soil culture samples ranged from 0 to 0.8, and the spectral curves showed dispersion in the 400–2450 nm band. There are significant absorption peaks at the 1400 nm, 1900 nm and 2200 nm bands, mainly associated with moisture, clay, and minerals in the soil. (3) Mathematical transformation of the original soil spectrum can effectively enhance the characteristic areas of the spectral curve, reveal some hidden spectral feature information, and highlight the reflection peaks and absorption valleys of the spectral curve. The combination of multiple data transformation forms can further highlight the characteristic regions of the spectral curve than one data transformation from alone, which can better demonstrate better advantages in the subsequent modeling process. (4) Among the four inversion models, the modeling set accuracy of RF is greater than that of MLR, PLSR and SVM models, and has better robustness. The fitting effect of the two machine-learning models, RF and SVM, is generally higher than that of MLR and PLSR, but not absolutely. Among the multiple data transformation forms, the model fitting effect and stability of the combined data transformation was more obvious, presumably because the combined data transformation form could tap into the information of the inverse spectral reflection characteristics of organic carbon in soil. Therefore, different mathematical transformation forms matched with different models produce different inversion effects, which can only be explored more intuitively in practical applications. (5) Among the four soil nutrient indicators, MSCFD-RF has higher prediction accuracy for SOC and TN. the simulation results of MSC-RF model for TK are better and can achieve the desired results. The RPD values of all inversion models of TP are lower than 1.4, and the simulation results are less satisfactory. However, the reliability of various finger data transformation forms for exploring spectral information can be effectively evaluated by comparing the accuracy of all spectral construction models.

In this study, the sampling sites were mainly concentrated in the northeastern and a few central parts of the TRSR region, with few western areas involved, and the uneven distribution of sampling sites. This may affect the accuracy of the model inversion. However, for the entire TRSR region, the number of sampling points needs to be increased and the sampling scheme needs to be optimized in subsequent studies. The scientific delineation of sampling points will make the study results more scientific. Secondly, the next step to carry out the work is mainly to combine with remote sensing images to verify and compare the inversion results with the research results of this paper. Enriching the soil spectral database to form a set of large-scale dynamic monitoring methods of soil nutrients is of great practical significance for soil quality protection.

References

Ben-Dor, E., Inbar, Y., & Chen, Y. (1997). The reflectance spectra of organic matter in the visible near-infrared and short wave infrared region (400–2500 nm) during a controlled decomposition process. *Remote Sensing of Environment, 61*, 1–15.

Chang, C., & Laird, D. A. (2002). Near-infrared reflectance spectroscopic analysis of soil c and n. *Soil Science, 167*, 110–116.

Dalal, R. C., & Henry, R. J. (1986). Simultaneous determination of moisture, organic carbon, and total nitrogen by near infrared reflectance spectrophotometry. *Soil Science Society of America Journal, 50*, 120–123.

Du, R. (2018). *Hyperspectral remote sensing estimation of soil nutrients in Alpine Grassland* (CNKI). Master thesis, Chengdu University of Technology.

Gao, H., Chen, H. Y., Liu, H. T., Tan, L. M., & Liu, J. T. (2013). Spontaneous determination of soil available phosphorus using high spectrum in the Northwest Plain of Shandong Province. *Chinese Journal of Eco-Agriculture, 21*, 752–757.

Huang, Y.F., & Liu, T. (1995). Spectral characteristics of main types of soils in southern China and soil classification. *Acta Pedologica Sinica, 1*, 58–68.

Ji, W., Li, X., Li, C.-X., Zhou, Y., & Shi, Z. (2012). Using different data mining algorithms to predict soil organic matter based on visible-near infrared spectroscopy. *Spectroscopy and Spectral Analysis, 32*, 2393–2398.

Liu, C., Li, X., Wang, W., Gao, Y., Zhou, Z., & Zhao, Z. (2011). Research on the detection method of total nitrogen in soil based on hyperspectral image technology. *Chinese Society of Agricultural Engineering 2011*, 1258–1261.

Lu, Y., Bai, Y., Yang, L., & Wang, H. (2007). Prediction and validation of soil organic matter content based on hyperspectrum. *Scientia Agricultura Sinica, 9*, 1989–1995.

Nan, F. (2017). *Hyperspectral inversion of major soil nutrients in reclaimed farmland in the coal mining area of Loess Plateau* (CNKI). Dissertation, Shanxi Agricultural University.

Peng, J., Zhang, Y.-Z., Zhou, Q., Liu, X.-L., & Zhou, W.-J. (2006). Spectral characteristics of soils in Hunan Province as affected by removal of soil organic matter. *Soils, 4*, 453–458.

Peng, Y., Zhang, J., He, X., & Lu, E. (1998). Analysis of soil moisture, organic matter and total nitrogen content in loess in china with near infrared spectroscopy. *Acta Pedologica Sinica, 4*, 553–559.

Qi, Y., Zhang, X., Zhang. L., Lv, X., Zhang, Z., Chen, J., Li, X., Wang, F., & Peng, K. (2018). Research on soil nutrient content estimated model by hyperspectral remote sensing data. *Xinjiang Agricultural Sciences, 55*, 490–495.

Stenberg, B. (2010). Effects of soil sample pretreatments and standardised rewetting as interacted with sand classes on vis-NIR predictions of clay and soil organic carbon. *Diffuse Reflectance Spectroscopy in Soil Science and Land Resource Assessment, 158*, 15–22.

Suk Lee, W., S. Mylavarapu, R., Seob Choe, J., & D. Whitney, J. (2001). *Study on soil properties and spectral characteristics in florida*. 2001 ASAE Annual Meeting, St. Joseph, MI.

Viscarra Rossel, R. A., & Webster, R. (2011). Discrimination of Australian soil horizons and classes from their visible–near infrared spectra. *European Journal of Soil Science, 62*, 637–647.

Xu, B., & Dai, C. (1980). Correlation analysis of spectral reflectance properties and organic matter content of soils in South Xinjiang. *Science Bulletin, 6*, 282–284.

Yang, Y. (2014). *Hyperspectral inversion of total nitrogen, total carbon and carbon to nitrogen ratio of soils in the TRSR region*. Master thesis, Qinghai Normal University.

Yu, F., Min, S., Ju, X., & Zhang, F. (2002). Determination the content of nitrogen and organic substance in dry soil by using near infrared diffusion reflectance spectroscopy. *Chinese Journal of Analysis Laboratory*, *3*, 49–51.

Yu, L., Hong, Y. S., Geng, L., Zhou, Y., Zhu, Q., Cao, J. J., & Nie, Y. (2015). Hyperspectral estimation of soil organic matter content based on partial least squares regression. *Transactions of the Chinese Society of Agricultural Engineering*, *31*, 103–109.

Yu, L., Wei, D., Wang, H., Huang, Q., Peng, Y., & Xu, Y. (2016). Spatial variability of soil organic matter and appropriate number of samples on county scale in Jianghan Plain. *Journal of Natural Resources*, *31*, 855–863.

Yuan, S., Ma, T., Song, T., He, Y., & Bao, Y. (2009). Real time analysis of soil total N and P with near infrared reflectance spectroscopy. *College of Biosystems Engineering and Food Science*, *40*, 150–153.

Zhang, C., Liu, Y.-M., Sun, Y.-N., Wang, L., & Liu, J.-H. (2018a). Hyperspectral prediction model of soil nutrient content in the loess hilly-gully region. *Chinese Journal of Applied Ecology*, *29*, 2835–2842.

Zhang, D., Zhao, Y., Qin, K., Zhao, N., & Yang, Y. (2018b). Influence of spectral transformation methods on nutrient content inversion accuracy by hyperspectral remote sensing in black soil. *Transactions of the Chinese Society of Agricultural Engineering*, *34*, 141–147.

Zhang, J., Tian, Y., Yao, X., Cao, W., Ma, X., & Zhu, Y. (2012). The spectral index for estimating soil OM, TN and AN content simultaneously using near-infrared spectroscopy. *Acta Pedologica Sinica*, *49*, 50–59.

Chapter 5
Grassland Degradation Evaluation Using Hyperspectral Remote Sensing

Abstract Grassland is an important natural resource; however, in recent years, due to the effects of climate warming and human disturbances, grassland degradation has become more severe. Consequently, the role of hyperspectral remote sensing in extracting ground object information and monitoring degradation has become increasingly significant. Taking Xinghai County as an example, this paper utilizes hyperspectral images from the China-Brazil Earth Resources Satellite (CBERS) 01, in conjunction with community survey data from sample plots. It extracts pixel-scale grassland information via spectral feature analysis and mixed spectral decomposition, performs grassland vegetation classification mapping, and classifies the level of grassland degradation based on the proportion of high-quality forage and vegetation. The main conclusions are as follows: (1) In analyzing the spectral characteristics of typical grassland vegetation, low-coverage grasslands can be identified through the reflectance of chlorophyll-sensitive bands (559 nm and 670 nm) in the visible light spectrum. In the near-infrared spectrum, the spectral curve exhibits significant fluctuations. Reflectance characteristics at 1122 nm and 1660 nm can be used to differentiate shady slope vegetation, linear-leaved Kobresia grassland communities, and crops from other types of vegetation. The first-order derivative transformation and envelope removal transformation enhance the spectral characteristic differences of various vegetation types in the "red edge" and "red valley" regions, respectively; (2) By employing the Mahalanobis distance method and correlation analysis, the reflectance spectra for 12 dominant grass species, including Alpine Songgrass, Dwarf Songgrass, and Wolf Poison Grass, were optimized and de-redundant, identifying 8 sensitive bands (507 nm, 670 nm, 713 nm, 765 nm, 1257 nm, 1324 nm, 1459 nm, 1929 nm) extracted. (3) Comparing the three hyperspectral remote sensing classification methods (spectral angle mapping, spectral information divergence and decision tree), it was found that the decision tree method based on spectral feature analysis had the highest overall classification accuracy (77.56%), and achieved ideal classification results at the grassland community level; (4) Multiple endmember spectral mixture analysis (MESMA) was used to study the mixed pixel decomposition of grassland communities, and the results showed that MESMA had good accuracy for spectral mixture decomposition between vegetation and bare soil (RMSE of 0.16), but poor performance for mixed decomposition between fine forage, weed grass,

© The Author(s), under exclusive license to Springer Nature Singapore Pte Ltd. 2026

W. Zhou et al., *Carbon Sink Estimation and Evaluation of Ecological Function Changes on the Tibetan Plateau*, https://doi.org/10.1007/978-981-95-8664-6_5

and bare soil (RMSE of 0.28); (5) Two evaluation indicators of grassland degradation, namely "vegetation ratio" and "excellent forage ratio," were proposed, and the grassland degradation was evaluated hierarchically, and the results showed that the proportion of moderate grassland degradation in the study area reached 56.25%.

Keywords Hyperspectral remote sensing · Grassland vegetation identification · Grassland degradation monitoring · Mixed pixel decomposition

5.1 Introduction

From an ecological standpoint, grassland degradation is a process of regressive succession within grassland ecosystems, influenced by human activities such as overgrazing and land reclamation, as well as adverse natural conditions. It represents one of the primary forms of land desertification. In terms of management, grassland degradation is characterized by a decrease in productivity, resulting from reduced forage yield and quality (Li, 1997). Given the regional variations, different grassland types, and the diverse causes of degradation, the characteristics and manifestations of grassland degradation are complex. The primary indicators for monitoring grassland degradation include grassland coverage, grass layer height, plant community structure, indicator plants, aboveground biomass, and soil nutrients (Su et al., 2003).

Compared to field quadrat surveys, remote sensing technology offers the advantages of a wide monitoring range and high timeliness in grassland monitoring. In particular, the development of hyperspectral remote sensing technology enables the extraction of spectral feature information from features with hundreds or thousands of bands, significantly improving the accuracy of vegetation spectral feature recognition and classification (Zhou et al., 2009). Since the 1930s, aerial images have been internationally utilized for resource surveys and have been widely applied in grassland monitoring and livestock management (Tueller, 1989). From the late twentieth century to the early twenty-first century, the study of vegetation spectral characteristics led to the discovery of various vegetation indices, including the Normalized Difference Vegetation Index (NDVI), (Purevdorj et al., 1998) as well as ecological indicators such as grassland coverage, biomass (Lobo et al., 1998), and biodiversity (Muldavin, 2001). After 2010, with the advancement of high spatial resolution and hyperspectral remote sensing, coupled with the continuous development of ecological and biochemical theories, grassland vegetation type identification (Marcinkowska-Ochtyra et al., 2018; Sandino et al., 2018), grassland chemical composition index extraction (Clevers & Kooistra, 2012), grassland utilization intensity assessment (Franke et al., 2012), and grassland ecosystem service assessment (Obermeier et al., 2019) have emerged as new areas of focus. Grassland resource remote sensing research has exhibited a trend toward greater refinement, comprehensiveness, and systematization.

Quantitative remote sensing of grassland vegetation is inverted from a spectral characteristic index model established on the basis of multispectral data (Zhang et al.,

2007), and has evolved to include modeling inversions based on hundreds of hyperspectral bands (Chen & Zhang, 2017). The modeling method has progressed from linear regression to machine learning (Yang et al., 2018), significantly improving inversion accuracy. Current applications primarily involve the estimation of vegetation and bare soil abundance, (Guerschman et al., 2009; Zhang et al., 2013) as well as the extraction of forage and weed abundance (Ai et al., 2020). In specific applications, Hyperion data and HSI data have been thoroughly validated. Higher spatiotemporal resolution data, such as the AHSI hyperspectral data carried by the ZY1E satellite, have been applied to classify coastal wetland cover in Hangzhou Bay, achieving a Kappa coefficient of 0.93 (Wei et al., 2020) and confirming the potential of this data for complex land cover classification. According to previous research (Jin et al., 2017; Wei et al., 2019; Yang et al., 2012), the reflection characteristics of grassland vegetation in the visible and near-infrared spectrum are mainly influenced by community height, aboveground biomass, coverage, leaf chlorophyll content, leaf area index, and community canopy structure.

However, due to factors such as spatial resolution, surface heterogeneity, and complex terrain, the issue of mixed pixels has always posed a technical challenge for remote sensing technology in the recognition and extraction of grassland information. Mixed pixel decomposition aims to establish a mixed model that characterizes the spectral composition within a pixel, a process known as spectral mixing decomposition. Ichoku (Ichoku & Karnieli, 1996) categorizes pixel mixing models into five types: linear models, probabilistic models, geometric optical models, stochastic geometric models, and fuzzy analysis models. The current linear model is typically used in the practical application of mixed pixel decomposition due to its simplicity in modeling and clear physical significance, whereas nonlinear demixing requires consideration of more complex influencing factors in both principle and implementation (Yang & Wang, 2017).

In actual decomposition, the linear mixed spectral decomposition method with a fixed number of endmembers has the drawback of either excessive endmember decomposition or too few, resulting in low decomposition accuracy. Roberts et al. (1998) proposed multiple endmember spectral mixture analysis (MESMA) based on variable endmember types and numbers. MESMA treats pixel spectra as a linear mixture of endmember spectra, using different ground object spectra to form multiple endmember models. It also allows for the selection of multiple spectra for each ground object and calculates the optimal endmember combination and abundance of pixels through optimal endmember judgment criteria, effectively addressing the problem of "same object, different spectra" in classification (Qi et al., 2016). Some studies have utilized mixed pixel decomposition or machine learning methods (Ai et al., 2020) to extract remote sensing indicators of grassland degradation based on the proportion of edible grass.

The extraction of indicators for evaluating grassland degradation is a key aspect of remote sensing monitoring of grassland health. The "Classification Indicators of Degradation, Desertification, and Salinization of Natural Grassland" (GB 197377-2003) established six categories for grassland degradation: plant community characteristics, community plant composition, degraded plants, aboveground biomass yield, surface characteristics, and soil physical and chemical properties. Xin and Du (2013) proposed four primary indicators of degradation for natural grasslands (in Qinghai Province—total vegetation coverage, plant species richness index, proportion of high-quality forage, and proportion of toxic and harmful plants—and identified 23 species as indicator plants for degraded grasslands. Vegetation degradation typically precedes soil degradation (Yan & Tang, 2008). In the process of extracting remote sensing indicators for grassland degradation assessment, some researchers opt for individual indicators such as biomass (Han et al., 2018), coverage, and net primary productivity (Jiang et al., 2007), while others employ multiple indicators for a comprehensive evaluation (Yao et al., 2016). Generally, the decline in coverage, aboveground biomass, and productivity are the most sensitive and significant indicators of grassland degradation, making the three indicators of aboveground biomass, vegetation coverage, and net primary productivity the most frequently used. Furthermore, the extraction of vegetation and non-vegetation abundance with the aid of spectral hybrid pixel decomposition technology can also characterize the vegetation coverage index to a certain extent (Guerschman et al., 2009; Zhang et al., 2013). Zhao et al. (2004) proposed evaluation indicators for toxic weed-type degraded grasslands. The separability of grassland vegetation in the spectrum has been thoroughly validated. Qu et al. (2013) introduced a remote sensing monitoring method using hyperspectral technology to detect the invasion of poisonous weeds. Wang et al. (2010) measured the leaf canopy and reflectance spectral data of various plants, including Leymus chinensis, Stipa cruzi, and Artemisia annua, in Hulunbuir, and decomposed the dominant species and degradation indicator species into mixed spectra. Their results indicated that the spectral recognition accuracy exceeded 95%.

The three rivers source region is situated in the southern part of Qinghai Province, within the heartland of the Tibetan Plateau. Covering an area of 302,500 km^2, 65.37% of this land is grassland. The region is predominantly an area of animal husbandry, where grassland serves as a vital natural resource, spanning 214,100 km^2 of natural grassland. However, the three rivers source region is also one of the most ecologically sensitive and fragile areas in China. Since the 1970s, the grasslands have experienced degradation due to a combination of natural and human factors, including climate change, rodent and pest infestations, and overgrazing. This degradation (Liu et al., 2008) poses a significant threat to the local ecological environment and the practice of animal husbandry. Following the implementation of the Three Rivers Source Region Ecological Protection and Construction Project after 2005, the overall grassland ecosystem has shown signs of improvement (Xu et al., 2017). In recent years, the warming and humidifying climate trends on the Qinghai-Tibet Plateau have further aided in the restoration of grassland ecosystems (Wei et al., 2019). Nevertheless, artificially treated "black soil type" degraded grasslands still face the risk of secondary degradation due to factors such as the invasion of poisonous weeds and a lack (Shang et al., 2018) of biodiversity. Overgrazing and rodent and insect infestations have also introduced uncertainty regarding the dominant species status within the grassland plant community structure. Zhao et al. (2005). Therefore, despite the overall trend of restoration in the grassland ecosystem of the three rivers source region, localized grassland degradation, characterized by changes in community composition and structure, persists. Long-term dynamic monitoring and detailed evaluation of grassland vegetation are crucial for the restoration of degraded grasslands.

In this study, spectral feature matching methods, including the spectral angle mapper (SAM), spectral information divergence (SID), and decision tree classification (DTC), were employed using hyperspectral data from the ZY1E satellite to classify the grassland communities at the community level.

5.2 Materials and Methods

5.2.1 Study Area

Xinghai County is located in the eastern part of the Tibetan Plateau (34° 48′–36° 14′ N, 99° 01′–100° 59′ E), and is located in the core area of the "Three Rivers Source" national ecological protection (Fig. 5.1). The total area of the county is 12,100 km^2, the width from east to west is 118.75 km, and the length from north to south is 159.5 km. The terrain of Xinghai County as a whole is high in the southwest and low in the northeast, while the Ora Mountains rise in the northwest, the average altitude of Xinghai County is 4300 m. The climate in Xinghai County is a plateau continental climate, with an average annual temperature of 2 °C and an average annual precipitation of 382.4 mm, long winter without summer, sufficient sunshine, strong radiation, poor heat conditions, precipitation is mainly concentrated in June to September. Xinghai County has a total land area of 12.18×10^4 km^2, and 83% of the total area is grassland, and 1.04 million livestock in 2017.

5.2.2 Data Sources

5.2.2.1 Field Survey

We conduct field survey in Xinghai County in autumn of 2019 and 2020, including collect spectral data and grassland community quadrat information of 47 sample sites. Among 47 sampling sites, the number of alpine grassland sites is 11, alpine meadow sites is 15, alpine meadow grassland sites is 4, temperate grassland sites is 15 and mountain meadow sites is 2. The sampling site are showed in Fig. 5.2.

In this study, the ASD FieldSpec4 spectroradiometer was utilized to collect vegetation spectra. The field survey methods were as follows: The sampling route was determined with a comprehensive consideration of traffic accessibility, ensuring it was distant from main roads, high-voltage lines, residential areas, and other potential interference factors. Additionally, the sampling encompassed five grassland types. At each site, hyperspectral data was collected within a 1 m × 1 m grassland community quadrat, as well as in pure grass seed type and bare soil quadrats. Spectral

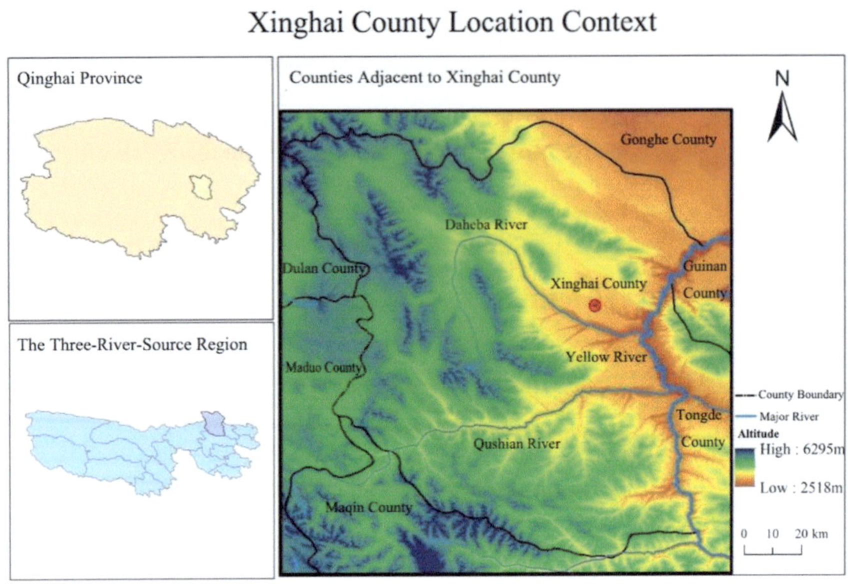

Fig. 5.1 Regional overview map of Xinghai County

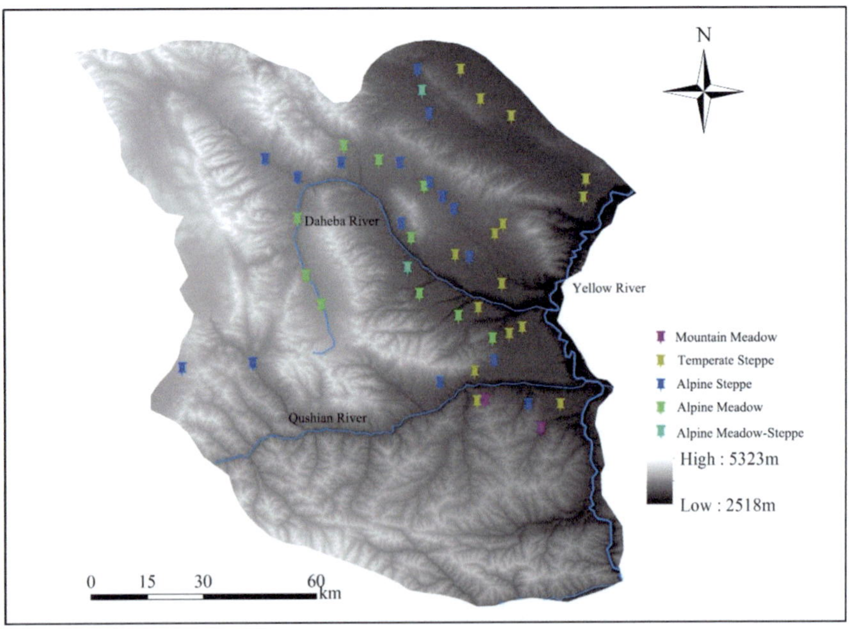

Fig. 5.2 Distribution of field sampling sites and grassland types

Table 5.1 ZY1E hyperspectral image information

Imaging time	Product number	Orbital line number	Orbital column number	Image range	Matrix of the number of cells
2020-09-15 04:25:39	151672	36	120	99.24° E–100.05° E 35.36° N–36.09° N	2051 * 1999

measurements were conducted under clear, cloudless weather conditions and with wind speeds below level 3. The measurement height was maintained at 1 m where possible, and for smaller plant populations, it was controlled at 0.5 m. All sampling points were precisely located using a handheld GPS. The soil properties, land use types, and vegetation cover conditions around the sampling points were recorded in detail. In total, spectral data from 47 vegetation sites were collected.

5.2.2.2 Hyperspectral Image Data and Its Acquisition

In this study, ZY1E-AHSI hyperspectral image data at the L1A level were utilized, which underwent radiation and system geometry corrections. The imaging time, range, and additional information are detailed in Table 5.1. The hyperspectral imaging data consists of a total of 166 spectral bands, with 76 bands in the visible near-infrared (VNIR) range from 395.86 to 1040.2 nm, featuring a spectral resolution of 10 nm. In the shortwave infrared (SWIR) range, from 1005.83 to 2501.1 nm, the spectral resolution is 20 nm.

5.2.3 Spectral Data Preprocessing

5.2.3.1 Field Measured Spectral Data

Firstly, the hyperspectral data collected in the field were loaded and screened using the ViewspecPro software of the ASD spectrometer. Abnormal or invalid data with large deviations, caused by soil environment and human operation, were eliminated, as well as the noise band. Next, the Savitzky-Golay smoothing algorithm (with a window width of 9 was applied to smooth the spectral data, effectively reducing noise while preserving the peak and trough characteristics. Finally, to match the band settings of ZY1E hyperspectral images, the Gaussian model resampling tool in ENVI software was used to complete the standardization of spectral data, which includes a total of 149 bands.

5.2.3.2 Hyperspectral Image Preprocessing

The preprocessing of ZY1E hyperspectral images involves image synthesis, radiometric calibration, atmospheric correction, and geometric precision correction, performed in sequence. This study removed four bad bands with wavelengths of 2450.88, 2467.57, 2484.32, and 2501.1 nm. The reason for their removal was the presence of noise in the ground-measured spectral data for these bands, and the 2501.1 nm band exhibited zero values in the image. Additionally, among the redundant bands, the first three bands of SW with lower spectral resolution (1005.6, 1022.21, and 1039.19 nm) were eliminated, and 10 bands, including the water vapor absorption bands of 1357.91–1408.49 nm and 1828.82–1912.97 nm, were excluded. The final synthesized image comprises 149 bands.

5.2.4 Methods

5.2.4.1 Grass Species Survey Statistics Analysis

Through the survey of grassland community quadrats and related knowledge, 31 species of herbaceous plants were identified as being mainly distributed in the study area. These plants belong to 10 families and 21 genera (Table 5.2). The dominant or co-dominant species primarily include six valuable forage grasses (*Krylov Needle-grass, Stipa purpurea Griseb, Achnatherum splendens, Kobresia pygmaea, Kobresia capillifolia, and Kobresia humilis*) and six harmful or undesirable weeds for cattle and sheep (*Stellera chamaejasme*, *Leontopodium*, Poa annua, *Oxytropis falcata*, Aster tataricus, and *Achnatherum inebrians*).

5.2.4.2 Classification of Grassland Community Types

In this study, we adhere to the grassland classification of the Natural Grassland Classification Standard (DB63/T 1513-2016) and the primary grass species of field samples. The standard defines a grassland type as a unit characterized by uniform habitat conditions, consistent plant communities, and similar utilization methods within a specific temporal and spatial range. Ultimately, seven grassland community types were delineated (Table 5.3).

Among them, the habitats of the two types of grassland communities, "*Stipa and Forbs*" and "*Achnatherum splendens and Stipa,*" exhibit clear zonal characteristics of temperate grasslands. These are primarily distributed below an altitude of 3500 m. The alpine meadows, including "*Kobresia pygmaea, forbs,*" "*Kobresia capillifolia,*" *and* "*Kobresia capillifolia, Polygonum viviparum,*" as well as the alpine grassland featuring "*Stipa purpurea Griseb, forbs,*" are located in regions of relatively high altitude. The main plants in these areas are shorter than those in temperate grasslands,

Table 5.2 Main grass species in the study area

Numbering	Latin name	Family	Genus	Classification by livestock value
1	*Krylov Needlegrass*	Gramineae	Needlegrass	Herbage
2	*Stipa purpurea Griseb*	Gramineae	Needlegrass	Herbage
3	*Poa annua*	Gramineae	Early maturing grass	Herbage
4	*Elymus nutans*	Gramineae	Genus alkali grass	Herbage
5	*Achnatherum splendens*	Gramineae	Genus Fern	Miscellaneous grass
6	*Achnatherum inebrians*	Gramineae	Genus Fern	Miscellaneous grass
7	*Leymus secalinus*	Gramineae	Rely on grasses	Miscellaneous grass
8	*Kobresia pygmaea*	Cyperaceae	Song grass	Herbage
9	*Kobresia capillifolia*	Cyperaceae	Song grass	Herbage
10	*Kobresia humilis*	Cyperaceae	Song grass	Herbage
11	*Carex spp*	Cyperaceae	Genus Grass	Herbage
12	*Stellera chamaejasme*	Ruixiangaceae	Wolf poison	Miscellaneous grass
13	*Leontopodium leontopodioides*	Compositae	Tinder grass	Miscellaneous grass
14	*Ajania tenuifolia*	Compositae	Asteracea	Miscellaneous grass
15	*Saussurea sauce*	Compositae	Genus Aegean aebrae	Miscellaneous grass
16	*Saussurea superba*	Compositae	Genus Aegean aebrae	Miscellaneous grass
17	*Ligularia virgaurea*	Compositae	Pack us	Miscellaneous grass
18	*Aster tataricus*	Compositae	Aster	Miscellaneous grass
19	*Artemisia sieversiana*	Compositae	Artemisia	Herbage
20	*Oxytropis falcata*	Leguminosae	Echinos	Miscellaneous grass
21	*Thermopsis lanceolata*	Leguminosae	Wild cassia is a genus	Miscellaneous grass
22	*Potentilla bifurca*	Rosaceae	Potentilla	Miscellaneous grass

(continued)

Table 5.2 (continued)

Numbering	Latin name	Family	Genus	Classification by livestock value
23	*Potentilla multifida*	Rosaceae	Potentilla	Miscellaneous grass
24	*Potentilla fruticosa*	Rosaceae	Potentilla	Miscellaneous grass
25	*Polygonum viviparum*	Polygonaceae	Polygonum	Miscellaneous grass
26	*Adenophora stenanthina*	Campanulaceae	Sand ginseng	Miscellaneous grass
27	*Gentiana scabra Bunge*	Gentianaceae	Gentiana	Miscellaneous grass
28	*Gentiana macrophylla*	Gentianaceae	Gentiana	Miscellaneous grass
29	*Pedicularis chinensis*	Scrophulariaceae	Horse first Artemisia genus	Miscellaneous grass
30	*Pedicularis spicata*	Scrophulariaceae	Horse first Artemisia genus	Miscellaneous grass
31	*Pedicularis alaschanica*	Scrophulariaceae	Horse first Artemisia genus	Miscellaneous grass

Table 5.3 Classification statistics of grassland community survey

Grassland community type	Sample sites	Altitude distribution (m)	Main grass species
Kobresia pygmaea, forbs	11	3245–4232	*Kobresia pygmaea, Kobresia humilis, Leontopodium leontopodioides, Stellera chamaejasme*
Krylov Needlegrass, forbs	11	3192–3488	*Krylov Needlegrass, Poa annua, Achnatherum splendens, Achnatherum inebrians*
Kobresia capillifolia, forbs	9	3232–3569	*Kobresia capillifolia, Stellera chamaejasme, Elymus nutans, Pedicularis chinensis*
Stipa purpurea Griseb, forbs	6	3375–3824	*Stipa purpurea Griseb, Poa annua, Leontopodium leontopodioides*
Achnatherum splendens, Krylov Needlegrass	6	3115–3452	*Achnatherum splendens, Krylov Needlegrass, Poa annua, Aster tataricus*
Kobresia humilis, Kobresia pygmaea	2	3331–3839	*Kobresia humilis, Kobresia pygmaea, Leontopodium leontopodioides*
Kobresia capillifolia, Polygonum viviparum	2	3360–3531	*Kobresia capillifolia, Kobresia capillifolia, Polygonum viviparum, Polygonum viviparum, Potentilla fruticosa*

with a higher plant density, smaller leaf area, and greater cold resistance. The difference between meadows and grasslands lies in the fact that grasslands are dominated by xerophytic herbaceous plants.

5.2.4.3 Spectral Characteristics Analysis of Typical Vegetation

In this study, two spectral transformation methods—spectral first-order derivative and envelope removal—were employed to analyze and identify the spectra of 9 typical vegetation samples for the spectral library. Spectral characteristic variables were *established to facilitate the development of the Decision Tree (DT) classification algorithm.*

5.2.4.4 Extraction of Sensitive Spectral Bands

In this study, the Mahalanobis distance method and correlation analysis were used to extract the sensitive bands. Compared with the Euclidean distance, the Mahalanobis distance is more suitable for high-dimensional data considering the relationship between various properties, and has the advantage of not being affected by scale and dimension.

$$MDk = \sqrt{(X-\mu)\mathrm{Cov}^{-1}(X-\mu)^{-1}} \tag{5.1}$$

where X is the reflectance difference matrix of the dominant grass species, μ is the zero-point vector, and Cov is the covariance matrix of X.

5.2.5 Vegetation Classification and Accuracy Verification

Common classification methods include spectral angle mapping (SAM), spectral information divergence (SID), and decision tree (DT), which are suitable for various data characteristics and application requirements. Once the classification is completed, a confusion matrix is used for accuracy verification, and overall accuracy (OA) and the Kappa coefficient are calculated to ensure the reliability of the classification results.

5.2.6 Multiple Endmember Spectral Mixture Analysis

In this study, we utilized multiple endmember spectral mixture analysis (MESMA) to detect and decompose mixed pixels. MESMA represents an enhancement over traditional spectral mixture analysis (SMA). It necessitates that the model adheres to constraints such as root mean square error, abundance values, and residuals, and permits a variable number and type of endmembers across pixels.

5.2.7 *Extraction of Degradation Indicators*

From the perspective of livestock use, excellent forage is over-gnawed, and the breeding of inedible or toxic and harmful weeds leads to a decrease in the proportion of edible forage for livestock, which is also a kind of grassland degradation in the sense of grassland management. In this study, these two characteristics were combined, and through the mixed spectral decomposition of the two levels, not only the vegetation abundance that could characterize the overall coverage, but also the excellent forage abundance and miscellaneous grass abundance that could characterize the grassland community structure composition were obtained. Based on the two, two remote sensing indexes, vegetation ratio (VegR) and good forage ratio (FineR), were established, so that the degradation degree of grassland vegetation could be evaluated at the scale of a single pixel (30 m). Vegetation ratio is the abundance of vegetation spectra in pixels, and FineR is calculated as follows:

$$\text{Fine}R = \frac{\text{Fraction(Fine)}}{\text{Fraction(Fine)} + \text{Fraction(Weed)}} \tag{5.2}$$

In the formula, Fraction (Fine) represents the abundance of excellent grass, while Fraction (Weed) represents the abundance of miscellaneous grass.

In the grading evaluation of grassland degradation, this study referred to the natural grassland degradation grading standards and evaluation indicators proposed by the Qinghai Provincial Grassland Central Station (Xin et al., 2013).

In this study, the total vegetation coverage and the proportion of good forage were characterized by VegR and FineR, respectively, and the grading criteria were defined by using a decision tree, with the total vegetation coverage as the main reference index and the proportion of good forage as the auxiliary. The grassland on the image is divided into four levels according to the degree of degradation: undegraded, mildly degraded, moderately degraded, and severely degraded.

5.3 Results

5.3.1 Spectral Characteristics Analysis of Typical Vegetation

As shown in Fig. 5.3, within the visible light range (380–780 nm), there is a reflection peak (i.e., the "green peak," which appears around 559 nm) and an absorption valley (i.e., the "red valley," which appears around 670 nm). These features respectively characterize the strong reflection and absorption spectral characteristics of chlorophyll in the green and red-light bands.

Due to the substantial decrease in the absorption rate of chlorophyll near the boundary between visible light and near-infrared and the sharp rise in vegetation reflectivity, the spectral characteristics of this wavelength range (i.e., the "red edge", 680–760 nm) are significantly correlated with the physicochemical characteristics of vegetation (leaf area index, chlorophyll a and chlorophyll b content per unit area, leaf structure characteristics, etc.).

In the near-infrared band (780–2526 nm), the reflectivity of vegetation is mainly determined by the internal structure of leaves and plant water content and is affected by multiple spectral absorption bands. Among these, the reflectivity around the wavelengths of 1.4 μm and 1.9 μm is severely disturbed by atmospheric moisture and aerosols, and the signal-to-noise ratio is low.

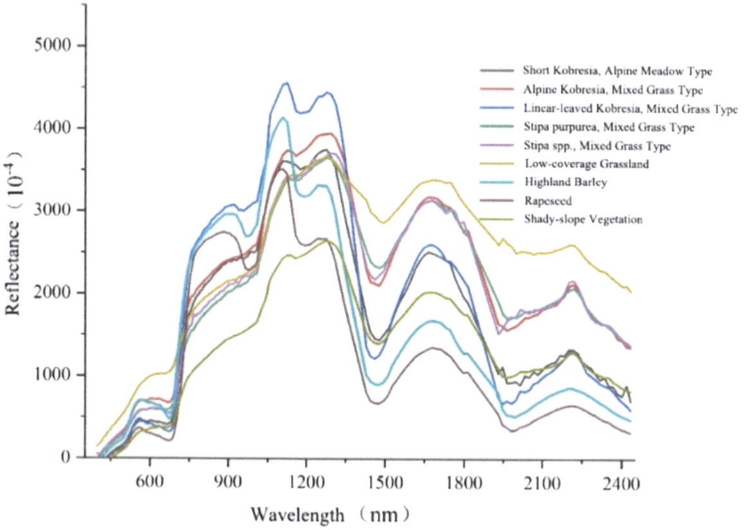

Fig. 5.3 Reflection spectral curves of typical vegetation. *Note* Type 1 is *Kobresia humilis, Kobresia pygmaea;* Type 2 *is Kobresia pygmaea, forbs;* Type 3 *Kobresia capillifolia, forbs*; Type 4 is *Stipa purpurea Griseb, forbs;* Type 5 is *Krylov Needlegrass, forbs;* Type 6 is low covered vegetation, Type 7 is highland barley, Type 8 is rape, and Type 9 is vegetation on shady slopes

Table 5.4 Reflectance of characteristic band of typical objects

Community types	Sensitive band reflection					
	R559	R670	R1122	R1274	R1660	R2199
Kobresia humilis, Kobresia pygmaea	0.0452	0.042	0.3611	0.3754	0.2512	0.1333
Kobresia pygmaea, forbs	0.0689	0.0677	0.3742	0.3945	0.3183	0.2105
Kobresia capillifolia, forbs	0.0481	0.0324	**0.4561**	0.4449	0.2598	0.1307
Stipa purpurea Griseb, forbs	0.0567	0.0594	0.3396	0.3658	0.3134	0.206
Krylov Needlegrass, forbs	0.0582	0.0555	0.3437	0.3683	0.3127	0.2161
Low coverage grassland	**0.0878**	**0.1045**	0.3418	0.3678	0.3378	0.2609
Highland barley	0.0706	0.0482	0.4086	0.3315	**0.1672**	0.0868
Rapeseed	0.0372	0.0214	0.346	0.2671	**0.1335**	0.0655
Mountain shadows	0.0313	0.0379	**0.2468**	0.2651	0.2024	0.1282

Bold indicates that the spectral reflectance of a certain type of grass is significantly higher or lower than that of other grass types. Therefore, by utilizing this difference, different grass species can be distinguished.

In the range of 780–1274 nm, due to the protection mechanism of leaves against burns and the influence of multiple narrow absorption bands of water and oxygen, the reflectivity of vegetation in this range is generally high, and the peak reflectance reaches its maximum at 1122 nm or 1274 nm.

In addition, two more obvious reflection peaks are formed at 1660 nm and 2199 nm. The reflectivity of vegetation at 1660 nm is mainly affected by the total water content of plants, and its reflectivity is inversely proportional to the water content. Meanwhile, around 2199 nm is the characteristic band of hydroxide groups, which is related to soil composition (mainly mineral composition).

As shown in Table 5.4, the spectral reflectance at peak and valley positions was extracted through the aforementioned spectral analysis, and the reflection characteristic variables of 9 types of vegetation were established. Among them, the reflectivity of "low-coverage grassland" at the Green Peak and Red Valley was high ($R_{559} = 0.878$, $R_{670} = 0.1045$), which could be distinguished from other types of vegetation. The reflectance of "Songgrass and Miscellaneous Grass Types" reached the reflectance peak at 1122 nm (0.4561), which was significantly higher than that of other vegetation types. In contrast, the reflectance of vegetation on the shady slope at 1122 nm (0.2468) was significantly lower than that of other vegetation types. The spectral characteristic variable R_{1660} could better distinguish the two crops (highland barley and rapeseed) from other vegetation types.

5.3.2 *Sensitive Spectral Band Identification of Main Grass*

According to the analysis of the sensitive bands extracted from the three data forms, the spectral reflectance has more sensitive bands in the green and red bands controlled by the vegetation chlorophyll content. The sensitive bands of the first—derivative transformation are mainly distributed near the red edge, as well as the reflection peak and absorption valley where the reflectivity changes rapidly. Meanwhile, the sensitive bands of the envelope removal transformation are mainly concentrated in the three intervals of 705–756, 902–945, and 2216–2417 nm.

In this study, the correlation coefficient and significance of the sensitive bands selected by paired Mahalanobis distances were analyzed using Origin software. By combining the correlation coefficient and T-test, the magnitude of the correlation between variables and their significance level were analyzed (Fig. 5.4).

Through correlation analysis, the optimized sensitive identification bands (i.e., the adjacent strong—correlation bands with $p < 0.001$ that passed the significance test after redundancy removal) between grass species were obtained under the three spectral data forms.

Among them, the optimized sensitive identification bands of the reflectance spectrum are 507, 670, 713, 765, 1257, 1354, 1459, and 1929 nm, a total of 8 bands.

The optimized sensitive identification bands of the first-derivative transformation spectrum are 542, 687, 1105, 1122, 1425, 1442, 1644, 1677, 1695, 1795, 1812, 1946, 2014, 2249, and 2417 nm, a total of 15 bands.

The optimized sensitive identification bands of the envelope removal transform spectrum are 447, 756, 816, 928, 1122, 1274, 2216, and 2417 nm, a total of 8 bands.

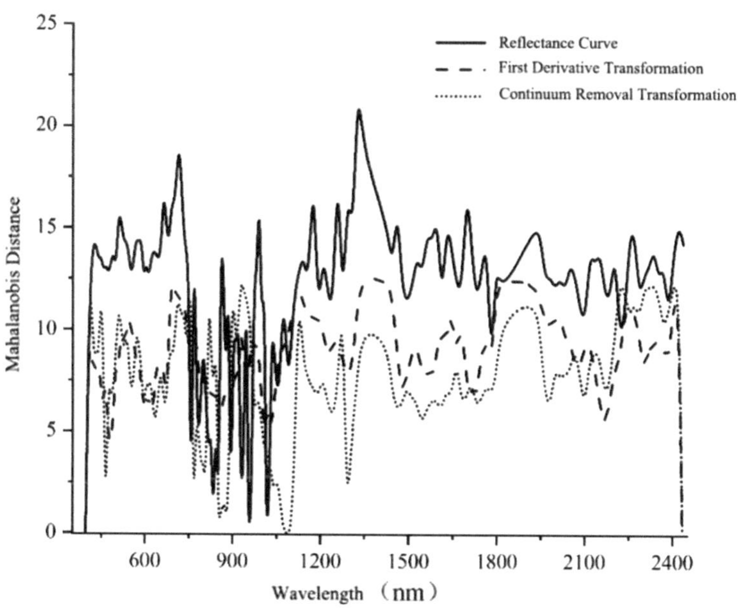

Fig. 5.4 Mahalanobis distances for each band for the three spectrum transformation

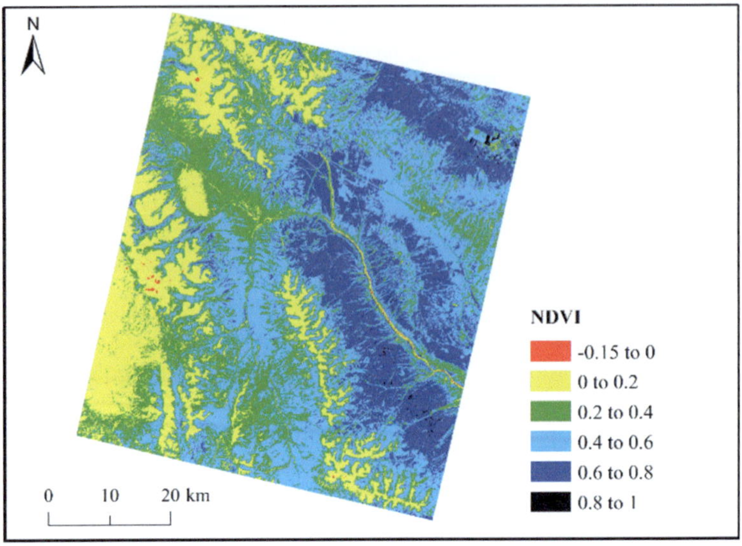

Fig. 5.5 NDVI image generated by calculation

5.3.3 *Binary Classification of Vegetation and Non-vegetation*

The NDVI image (Fig. 5.5) was calculated. The NDVI value range was [− 0.15, 1], and it was mainly distributed between 0 and 0.8, showing a pattern of high in the east and low in the west. Among them, about 12.77% of the study area had NDVI values between 0 and 0.2, 14.3% of the study area had NDVI values between 0.2 and 0.4, 27.8% of the study area had NDVI values between 0.4 and 0.6, and 13.78% of the study area had NDVI values between 0.6 and 0.8.

According to the original hyperspectral image, several points of interest were selected for statistical analysis of the NDVI of typical cover types (Table 5.5). It can be seen that the NDVI of non-vegetation pixels is basically less than 0.22. While only the impervious surface (buildings and roads), which has a low proportion in the image, has a slightly higher NDVI than this value, but it has little effect on the overall classification accuracy. So, 0.22 is set as the segmentation threshold between vegetation and non-vegetation, and the vegetation part with an NDVI greater than 0.22 is extracted (Fig. 5.5).

5.3.4 *Comparison of Classification Accuracy*

In this study, 50 verification points (each point occupies one pixel) were selected for each vegetation type based on the grassland type map of Qinghai Province, and then the accuracy of the classification results was verified by using a confusion matrix, and

Table 5.5 NDVI statistics of typical objects in images

Covering type	Average value	Numerical distribution
Vegetation	0.57	0.22–0.85
Bare soil	0.1	0.03–0.22
Cloud	0.11	− 0.03 to 0.19
Water body	0.01	− 0.01 to 0.06
Frozen soil	0.12	0.04–0.17
Impervious surface	0.13	0–0.35

the overall accuracy (OA) and Kappa coefficient of mapping, as well as the mapping accuracy of various types of vegetation, were calculated.

Through classification accuracy comparison analysis, decision tree classification accuracy was highest, overall accuracy was 77.56% versus 75.33% (spectral angle filling mapping) versus 72.89% (spectral information divergence). Therefore, here only decision tree classification process was showed.

The prepossessed hyperspectral images were input into the classifier to obtain a grassland vegetation classification map with 2174 rows, 1989 columns, and a spatial resolution of 30 m (Fig. 5.6). The vegetation area accounted for about 79.7% of the image area, of which "*Kobresia pygmaea, forbs*" had the largest area (about 18.48% of the image area), followed by low-coverage grassland (17.26%), "*Stipa purpurea Griseb, forbs*" (15.42%), "*Kobresia humilis, Kobresia pygmaea*" (8.82%), shady slope vegetation (8.78%), "*Krylov Needlegrass, forbs*" (8.19%), "*Kobresia capillifolia, forb*s" (2.6%), rapeseed (0.13%), highland barley (0.03%).

After the confusion matrix verification (Table 5.6), the overall classification accuracy was 77.6%. Among the various types of grassland vegetation, the mapping accuracy of shady slope vegetation (64%) was relatively low. Specifically, 24% of the verification sampling sites were misclassified as "*Stipa purpurea Griseb, forbs*", and 8% of the verification points were misclassified as "*Krylov Needlegrass, forbs*". In contrast, the mapping accuracy of other vegetation types was over 70%.

The Kappa coefficient of the DT method was 0.7475, indicating that the classification was highly consistent.

Among the classification results, "*Kobresia pygmaea, forbs*" were the most widely distributed, and had a large altitude span. Non-vegetated areas and low-cover grasslands are mainly distributed in the western part of the image (Fig. 5.6). Their topography is mainly manifested as snow-capped mountain frozen soil and river valley fading zone.

The vegetation distribution on shady slopes was similar to that of "*Stipa purpurea Griseb, forbs.*" Its main vegetation composition consisted of cold-resistant plants, which were mainly distributed on the steep slopes of mountains at higher altitudes.

The main grass species of "*Krylov Needlegrass, forbs*" are heat-loving and drought-tolerant temperate plants. Therefore, they are mainly distributed in the northeast corner of the area shown in the image, where the altitude is low and the rainfall is scarce.

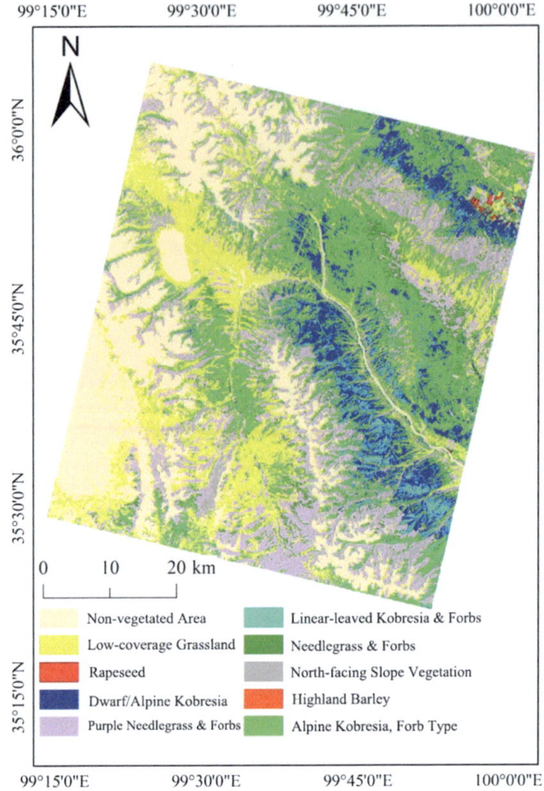

Fig. 5.6 Grassland vegetation classification map based on decision tree classification (DT)

The grassland communities of "*Kobresia capillifolia, forbs*" are characterized by high vegetation density and are mainly distributed in plains or gentle slopes at an altitude of 3000–3500 m.

Rape and highland barley are concentrated in the northeast of the study area, which has gentle slopes, low altitude, moderate heat, and is suitable for crop planting.

5.3.5 Analysis of Mixed Pixel Decomposition Results

In this study, the optimized end-member combination model and hyperspectral image are used as inputs. The value of the abundance of a single end-member is set to range from 0 to 1, and the sum of the abundances of each end-member being 1 is taken as the constraint to conduct multi-end-member mixed pixel decomposition.

The decomposition sequence is as follows: First, fit a simple two-end-member model and calculate the fitting error. After the error reaches the threshold, transfer the pixel to a more complex end-member model (three-end-member) for spectral

Table 5.6 Decision tree classification confusion matrix

	Type1	Type2	Type3	Type4	Type 5	Type6	Type7	Type8	Type9	Classification total
Type1	38	11	9	0	0	0	4	0	2	64
Type2	6	35	1	4	12	0	3	0	0	61
Type3	1	1	40	2	0	0	1	2	0	47
Type4	0	0	0	35	4	1	0	1	12	53
Type 5	0	1	0	4	34	0	0	2	4	45
Type6	5	2	0	5	0	49	0	0	0	61
Type7	0	0	0	0	0	0	41	0	0	41
Type8	0	0	0	0	0	0	1	45	0	46
Type9	0	0	0	0	0	0	0	0	32	32
Measure total	50	50	50	50	50	50	50	50	50	450
Drawing accuracy (%)	76	70	80	70	68	98	82	90	64	77.6

Note Type 1 is *Kobresia humilis, Kobresia pygmaea*; Type 2 *is Kobresia pygmaea, forbs*; Type 3 *Kobresia capillifolia, forbs*; Type 4 is *Stipa purpurea Griseb, forbs*; Type 5 is *Krylov Needlegrass, forbs*; Type 6 is low covered vegetation, Type 7 is highland barley, Type 8 is rape, and Type 9 is vegetation on shady slopes

fitting until the constraints are met. Finally, transmit the proportion of each end-member spectrum, that is, the abundance of end-members (Fig. 5.7). Here, L1 is the mixed decomposition result of pixels of vegetation and bare soil, and L2 is the mixed decomposition result of excellent forage, weed grass, and bare soil.

The accuracy of the two decomposition results was verified by the abundance obtained by mixed decomposition, compared with the vegetation coverage and separation coverage observed in the sample site (Table 5.7), and measured by root mean square error (RMSE).

The RMSE of the fitting value of vegetation coverage and the measured value was calculated to be 0.16, and the RMSE of the fitting value of the dominant forage coverage and the measured value was 0.28.

5.3.6 Grassland Degradation Indicators Extraction

The decrease in vegetation coverage is the most prominent feature of degraded grasslands, yet this is not comprehensive. From the perspective of livestock utilization, the over-grazing of high-quality grasses and the growth of inedible or toxic and harmful weeds result in a reduction in the proportion of grass that livestock can consume. This is also a form of grassland degradation in the context of grassland management.

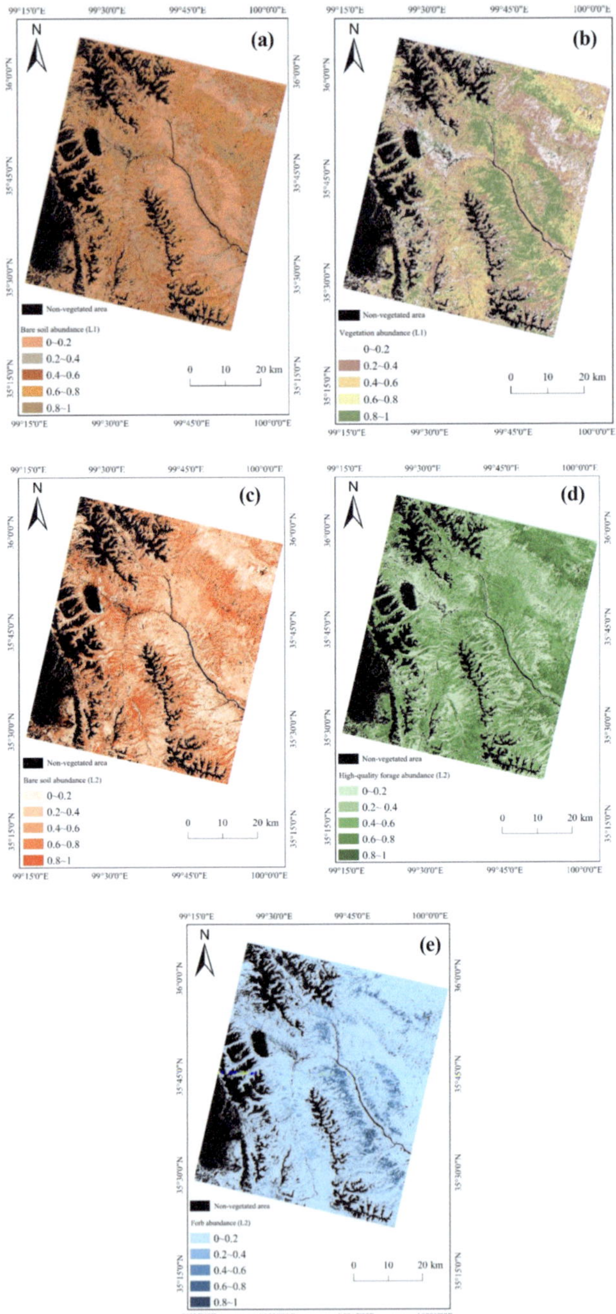

Fig. 5.7 Mixed pixels decomposition results (**a** and **b** are the abundance maps of each end element of "vegetation + bare soil" level decomposition, and **c**–**e** are the end element abundance maps of "excellent forage + weed grass + bare soil" level decomposition)

Table 5.7 Verifies the measured and MESMA fitting values of the sample points

Sample number	Vegetation abundance	Actual measurement coverage	Excellent forage abundance	Measured fractional coverage
190101	0.8156	0.9	0.6364	0.7
190903	0.4721	0.6	0	0.5
191302	0.6385	0.7	0.9927	0.65
191401	0.9589	0.9	0	0.1
200101	0.8481	0.9	0.7109	0.75
200202	0.2824	0.55	0	0.3
200508	0.8489	0.85	0.747	0.5
200801	0.9366	0.85	0	0.1
200901	0.9833	0.99	0	0.2
201001	0.455	0.65	0.7355	0.65
202104	0.2528	0.4	0.357	0.3
202201	0.888	0.9	0	0.6
202301	0.5729	0.75	0.5821	0.5
202401	0.5062	0.4	0.5095	0.25
202501	0	0.3	0.2579	0
202801	0.4011	0.65	0.5136	0.5
202901	0.5889	0.85	0.9176	0.4

In the grading evaluation of grassland degradation, this study referred to the natural grassland degradation grading standards and evaluation indicators proposed by the Qinghai Provincial Grassland General Station (Xin et al., 2013). In the original text, the quantitative indicators for classifying degraded grassland levels are as follows:

(1) For lightly degraded grassland: The total vegetation coverage of lightly degraded grassland is 70–80%, the plant species richness index is 14–25, the proportion of excellent grass is 35–70%, and the proportion of toxic, harmful, and inedible grass is 12–33% (or less than 33%).
(2) For moderately degraded grassland: The total vegetation coverage of moderately degraded grassland is 40–70%, the plant species richness index is 7–14, the proportion of excellent grass is 8% to 35%, and the proportion of toxic, harmful, and inedible grass is 33–67%.
(3) For severely degraded grassland: The total vegetation coverage of severely degraded grassland is less than 40%, the species richness index is less than 7 species, the proportion of excellent grass is less than 8%, and the proportion of toxic, harmful, and inedible grass is greater than 67%.

This study simplifies the four aforementioned indicators into two categories: vegetation ratio and excellent grass ratio, denoted by VegR and FineR, respectively. It then implements the classification algorithm using a decision-tree classifier.

The main principle and logic are as follows: first, determine the degree of degradation based on the threshold of the vegetation ratio. Then, combine this with the threshold of the excellent grass ratio for a second judgment. The result of the second judgment will not be better than the first rating. That is, the vegetation ratio serves as the primary reference indicator, while the excellent grass ratio acts as an auxiliary one.

Classify the grassland on the image into four levels according to the degree of degradation: non-degraded, mildly degraded, moderately degraded, and severely degraded (Fig. 5.8).

According to the results of the grading evaluation, 9.63% of the grassland was non-degraded, 1.49% was mildly degraded, 56.25% was moderately degraded, and 32.63% was severely degraded.

5.4 Conclusions

This paper refers to the existing local standards of Qinghai Province, combined with traditional field surveys, hyperspectral remote sensing data and remote sensing data process methods, taking Xinghai County as an example, the grassland vegetation classification method and grassland degradation index extraction in typical pastoral areas are studied, and the main research conclusions are as follows:

(1) The spectral characteristics and separability of typical grassland vegetation were analyzed. In the visible light band, low-cover grassland could be identified through the reflectance of the chlorophyll-sensitive bands (559 nm and 670 nm). In the near-infrared band, the spectral curves of each vegetation fluctuated significantly, and the reflection characteristics at 1122 nm and 1660 nm could distinguish shady-slope vegetation, *Kobresia capillifolia* type grassland communities, and crops from other vegetation.

    The first-order derivative transformation and the envelope removal transformation can enhance the spectral characteristics of various types of vegetation in the "red edge" and "red valley" respectively. Thus, three characteristic variables, namely red edge slope, red edge area, and band depth, can be established to further identify the remaining vegetation.

(2) Combined with the Mahalanobis distance method and correlation analysis, the sensitive identification bands of 12 dominant grass species, such as *Kobresia pygmaea, Kobresia humilis, and Stellera chamaejasme*, were optimized, and the sensitive band of the reflection spectrum were 507 nm, 670 nm, 713 nm, 765 nm, 1257 nm, 1324 nm, 1459 nm, and 1929 nm. The sensitive recognition bands of the first-order derivative transformation are 542 nm, 687 nm, 1105 nm, 1122 nm, 1425 nm, 1442 nm, 1644 nm, 1677 nm, 1695 nm, 1795 nm, 1812 nm, 1946 nm, 2014 nm, 2249 nm, and 2417 nm. The sensitive recognition bands of the envelope removal transform spectrum are 447 nm, 756 nm, 816 nm, 928 nm, 1122 nm, 1274 nm, 2216 nm, and 2417 nm.

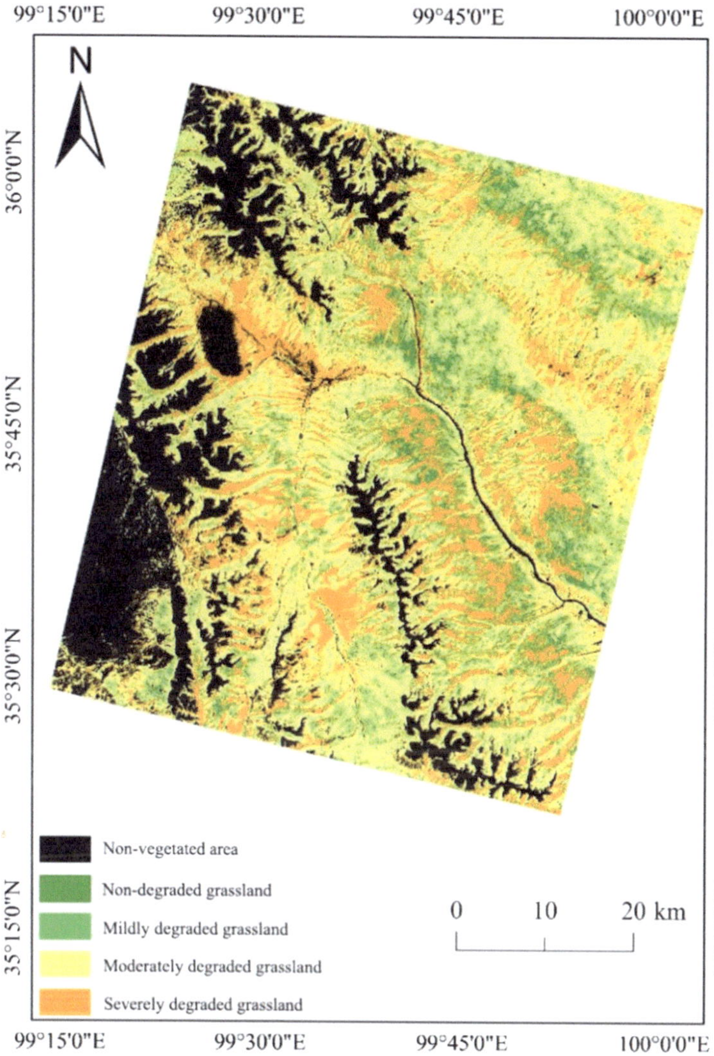

Fig. 5.8 Grading evaluation map of grassland degradation degree

(3) Using the ZY1E-AHSI image data, the three commonly used classification methods of hyperspectral remote sensing, spectral angle mapping, spectral information divergence and decision tree, and the total classification accuracy ranged from 72.9% to 77.6%, among which the decision tree classification method based on the analysis results of spectral characteristics had the highest accuracy, reaching 77.6%.

(4) Multi-end spectral mixed analysis (MESMA) was used to study the spectral demixing of features at different classification levels, and the results showed that MESMA had good accuracy for mixing and decomposition between vegetation and bare soil, but better data and fitting methods were still needed for mixed decomposition between grassland vegetation.
(5) Two grassland degradation indexes of total vegetation coverage and the proportion of excellent forage were calculated by the sub-pixel information of vegetation abundance, excellent forage and forbs abundance. The grassland degradation evaluation was carried out, and the results showed that 52.7% of study area reached to moderate degradation.

References

Ai, Z., An, R., Lu, C., & Chen, Y. (2020). Mapping of native plant species and noxious weeds to investigate grassland degradation in the Three-River Headwaters region using HJ-1A/HSI imagery. *International Journal of Remote Sensing, 41*, 1813–1838.

Chen, J., & Zhang, L. (2017). Estimating grassland coverage based on hyperspectral remote sensing in the northern Tian shan Mountain. *Prataculturai Science, 34*, 30–39.

Clevers, J. G. P. W., & Kooistra, L. (2012). Using hyperspectral remote sensing data for retrieving canopy chlorophyll and nitrogen content. *IEEE Journal of Selected Topics in Applied Earth Observations and Remote Sensing, 5*, 574–583.

Franke, J., Keuck, V., & Siegert, F. (2012). Assessment of grassland use intensity by remote sensing to support conservation schemes. *Journal for Nature Conservation, 20*, 125–134.

Guerschman, J. P., Hill, M. J., Renzullo, L. J., Barrett, D. J., Marks, A. S., & Botha, E. J. (2009). Estimating fractional cover of photosynthetic vegetation, non-photosynthetic vegetation and bare soil in the Australian tropical savanna region upscaling the EO-1 Hyperion and MODIS sensors. *Remote Sensing of Environment, 113*, 928–945.

Han, D., Wang, G., Xue, B., Liu, T., A, Y., & Xu, X. (2018). Evaluation of semiarid grassland degradation in North China from multiple perspectives. *Ecological Engineering, 112*, 41–50.

Ichoku, C., & Karnieli, A. (1996). A review of mixture modeling techniques for sub-pixel land cover estimation. *Remote Sensing Reviews, 13*, 161–186.

Jiang, L. P., Qin, Z. H., & Xie, W. (2007). A research of monitoring grassland degradation based on mono temporal MODIS data. *Chinese Journal of Grassland, 29*, 39–43.

Jin, G. L., Wei, X. H., Fan, Y. M., An, S. Z., Zhu, X. W., & Wang, H. N. (2017). Correlation between community characteristics and spectral reflectance of the ground on degraded Seriphidium transiliense Desert Grassland. *Acta Agrestia Sinica, 25*, 282–289.

Li, B. (1997). Grassland degradation in northern China and its prevention and control measures. *Scientia Agricultura Sinica, 30*, 1–10.

Liu, J. Y., Xu, X. L., & Shao, Q. Q. (2008). The spatial and temporal characteristics of grassland degradation in the Three-River Headwaters Region in Qinghai Province. *Acta Geographica Sinica, 63*, 364–376.

Lobo, A., Moloney, K., Chic, O., & Chiariello, N. (1998). Analysis of fine-scale spatial pattern of a grassland from remotely-sensed imagery and field collected data. *Landscape Ecology, 13*, 111–131.

Marcinkowska-Ochtyra, A., Jarocińska, A., Bzdęga, K., & Tokarska-Guzik, B. (2018). Classification of expansive grassland species in different growth stages based on hyperspectral and LiDAR data. *Remote Sensing, 10*.

Muldavin, E. H., Neville, P., & Harper, G. (2001). Indices of Grassland biodiversity in the Chihuahuan Desert Ecoregion Derived from Remote Sensing. *Conservation Biology, 15*, 844–855.
Obermeier, W. A., Lehnert, L. W., Pohl, M. J., Makowski Gianonni, S., Silva, B., Seibert, R., Laser, H., Moser, G., Müller, C., Luterbacher, J., & Bendix, J. (2019). Grassland ecosystem services in a changing environment: The potential of hyperspectral monitoring. *Remote Sensing of Environment, 232*, Article 111273.
Purevdorj, T., Tateishi, R., Ishiyama, T., & Honda, Y. (1998). Relationships between percent vegetation cover and vegetation indices. *International Journal of Remote Sensing, 19*, 3519–3535.
Qi, W. C., Zhang, X., & Yue, Y. M. (2016). Review on multiple endmember spectral mixture analysis. *Remote Sensing Information, 31*, 11–18.
Qu, R., Li, S., Xu, X. L., Wang, C. Z., & Tonglaga (2013). Progress on methods of Grassland degradation and weed invasion monitoring based on remote sensing. *Journal of Geo-Information Science, 15*, 761–767.
Roberts, D. A., Gardner, M., Church, R., Ustin, S., Scheer, G., & Green, R. O. (1998). Mapping Chaparral in the Santa Monica mountains using multiple endmember spectral mixture models. *Remote Sensing of Environment, 65*, 267–279.
Sandino, J., Gonzalez, F., Mengersen, K., & Gaston, K. J. (2018). UAVs and machine learning revolutionising invasive grass and vegetation surveys in remote arid lands. *Sensors (Basel, Switzerland), 18*, E605.
Shang, Z. H., Dong, Q. M., Shi, J. J., Zhou, H. K., Dong, S. K., Shao, X. Q., Li, S. X., Wang, Y. L., Ma, Y. S., Ding, L. M., Cao, G. M., & Long, R. J. (2018). Research progress in recent ten years of ecological restoration for 'Black Soil Land' degraded Grassland on Tibetan Plateau—— Concurrently discuss of ecological restoration in Sangjiangyuan Region. *Acta Agrestia Sinica, 26*, 1–21.
Su, D. X., Zhang, Z. H., & Chen, Z. Z. (2003). Discussion on the grading indicators of natural grassland degradation, desertification, and salinization in China: research on China's desertification control and industry—Celebrating the 10th Anniversary of the establishment of the China desertification control and industry society (1993–2003) Academic Papers Collection.
Tueller, P. T. (1989). Remote sensing technology for rangeland management applications. *Journal of Range Management, 42*, 442–453.
Wang, H. J., Fan, W. J., Cui, Y. K., Zhou, L., Yan, B. Y., Wu, D. H., & Xu, X. R. (2010). Hyperspectral remote sensing monitoring of Grassland degradation. *Spectroscopy and Spectral Analysis, 30*, 2734–2738.
Wei, D. D., Zhao, S. H., Xiao, C. C., Cui, H., & Liu, S. H. (2020). Leaf area index estimation algorithm of ZY-1-02D satellite. *Spacecraft Engineering, 29*, 169–173.
Wei, Y. Q., Lu, H. Y., Wang, J. N., Sun, J., & Wang, X. F. (2019). Responses of vegetation zones, in the Qinghai-Tibetan Plateau, to climate change and anthropogenic influences over the last 35 years. *Prataculturał Science, 36*, 1163–1176.
Xin, Y. C., & Du, T. Y. (2013). A Preliminary study on indexes of natural Grassland degradation classification in Qinghai. *Qinghai Prataculture, 22*, 19–21.
Xu, X. L., Wang, L., Li, J., & Cai, H. Y. (2017). Analysis of the grassland restoration trend and degradation situation in the "Three-River Headwaters" region since the implementation of the ecological project. *Journal of Geo-Information Science, 19*, 50–58.
Yan, Y. C., & Tang, H. P. (2008). Differentiation of related concepts of grassland degradation. *Acta Prataculturae Sinica, 17*, 93–99.
Yang, B., & Wang, B. (2017). Review of nonlinear unmixing for hyperspectral remote sensing imagery. *Journal of Infrared and Millimeter Waves, 36*, 173.
Yang, H. F., Li, J. L., Mu, S. J., Yang, Q., Hu, X. X., Jin, G. P., & Zhao, W. Y. (2012). Analysis of hyperspectral reflectance characteristics of three main grassland types in Xinjiang. *Acta Prataculturae Sinica, 21*, 258–266.

Yang, S., Feng, Q., Liang, T., Liu, B., Zhang, W., & Xie, H. (2018). Modeling grassland above-ground biomass based on artificial neural network and remote sensing in the Three-River Headwaters Region. *Remote Sensing of Environment, 204*, 448–455.

Yao, Z., Zhao, C., Yang, K., Liu, W., Li, Y., You, J., & Xiao, J. (2016). Alpine grassland degradation in the Qilian Mountains, China—A case study in Damaying Grassland. *Catena, 137*, 494–500.

Zhang, X., Liao, C., Li, J., & Sun, Q. (2013). Fractional vegetation cover estimation in arid and semi-arid environments using HJ-1 satellite hyperspectral data. *International Journal of Applied Earth Observation and Geoinformation, 21*, 506–512.

Zhang, Y. X., Zhang, Y. F., & Li, X. B. (2007). The synthetically estimating vegetation fractional coverage of grassland using field data and ASTER remote sensing imagine. *Chinese Journal of Ecology, 27*, 964–976.

Zhao, C. Z., Fan, S. Y., Yin, C. Q., & He, X. B. (2004). Study on vegetation community s structure of degraded grassland of noxious and miscellaneous grass type. *Journal of Desert Research, 24*, 507–512.

Zhao, H. X., Zhu, Z. L., Zhang, Y. L., Huang, X. Q., & Liu, J. (2005). Progress of research in mixed-seeding meadow. *Chinese Agricultural Science Bulletin, 26*(38–41), 49.

Zhou, L., Xin, X. P., Li, G., Yang, G. X., & Zhang, H. B. (2009). Application progress on hyperspectral remote sensing in grassland monitoring. *Prataculturał Science, 26*, 20–27.

Chapter 6
Estimation and Quantitative Assessment of Net Primary Productivity from 1982 to 2020

Abstract Changes in the vegetation net primary productivity (NPP) are primarily influenced by climate change and human activities. However, the dominant factors and their contributions at different time stages remain unclear. Here, we developed a method to quantify the relative contributions of climate and human factors to the vegetation NPP in China from 1982 to 2020 using three types of NPP components: the actual NPP (aNPP), potential NPP (pNPP), and human-occupied NPP (hNPP). The results showed that: (1) From 1982 to 2000, the annual average aNPP in China was 501.0 g C/m^2, with a total annual aNPP of 378.2 Tg C. From 2001 to 2020, the annual average aNPP increased to 517.1 g C/m^2, with a total annual aNPP of 391.1 Tg C. (2) From 1982 to 2000, human activities primarily drove aNPP changes, with contribution rates of 34.26% and 50.43% for increases and decreases, respectively. However, from 2001 to 2020, climate played a larger role in aNPP increases (the contribution was 43.90%), while human impact dominated decreases (the contribution was 83.75%). This study demonstrated that climate has played an increasingly important role in NPP increase over the past 40 years and will provide scientific strategies for climate adaptation.

Keywords Net primary productivity · Driving contribution · Residual method · Climate warming

6.1 Introduction

In recent decades, investigating the dynamics of ecosystem carbon sinks and quantifying the contributions of their driving factors have emerged as a scientific pot in climate change research. Among them, net primary productivity (NPP) is an important indicator of ecosystem functionality and plays an important role in the global carbon cycle (Rafique et al., 2016).

The impacts of climate change and human activities pose a threat to NPP stability, leading to significant changes in the structure, function, and overall ecosystem services of terrestrial ecosystems (Wang et al., 2020). Climate factors can directly

© The Author(s), under exclusive license to Springer Nature Singapore Pte Ltd. 2026
W. Zhou et al., *Carbon Sink Estimation and Evaluation of Ecological Function Changes on the Tibetan Plateau*, https://doi.org/10.1007/978-981-95-8664-6_6

or indirectly affect the NPP by regulating plant metabolism, growth season length, community biomass, etc. For example, from 1982 to 1999, the global NPP increased by 6%, particularly in tropical ecosystems, primarily due to a reduced cloud cover and increased solar radiation (Nemani et al., 2003). A severe drought and heat wave in Europe in 2003 caused a 30% reduction in the GPP and reversed four years of net ecosystem carbon sequestration (Ciais et al., 2005).

In regard to human activity, the impact on the NPP is double-sided. Among, land use/cover change as the most significant human activity, has exacerbated impacts on terrestrial ecosystems and vegetation productivity (Wang et al., 2018). On the one hand, urbanization and overgrazing can lead to a decrease in the NPP (Wu et al., 2020). On the other hand, ecosystem restoration projects such as afforestation or the return grazing land to grassland can promote an increase in the vegetation NPP (Sha et al., 2022). Our previous study found that from 1982 to 2010, the contributions of climate and human activities to grassland NPP decrease are nearly equal (48% vs. 46%), but for grassland NPP increase, human activities are the dominant factor, contributing 78.1% (Zhou et al., 2017). Meanwhile, the contributions of climate change and human activities to NPP changes vary over time. For example, in Qinghai-Tibet Plateau, climate was the dominant factor for the aNPP net increase, and its contribution rate dramatically increased from 36.3% during 1982–2000 to 224.2% during 2001–2020; Human activities changed from a positive contribution (53.3%) to a negative contribution (− 124.8%) to the aNPP net increase (Zhou et al., 2024).

Regarding changes in the vegetation NPP in China, studies have shown an increasing trend from 1982 to 1999 (Piao et al., 2005). And from 1982 to 2015, the annual NPP in China also showed a significant increase trend, particularly in the eastern and southwestern regions, mainly due to climate warming and national ecological restoration projects (Feng et al., 2019). Because, since the 1980s, China has implemented ecological projects such as returning farmland to forests and grasslands, desertification control, slope farmland control, in China, the positive effects of these projects have been recognized by the international community, such as promoting vegetation restoration and making China greener (Macias-Fauria, 2018; Piao et al., 2020), and southwest China has huge carbon sink under the ecological project (Tong et al., 2018).

Overall, some studies have quantitatively accounting the contribution of the climate and human activities to the NPP, but still exist several limitations. First, only the overall contribution rate of climate or human activities to NPP change has been quantitatively calculated, whereas the relative importance of specific factors related to NPP variation, such as climate, terrain, soil, and human management strategies, has been ignored. Second, land use/land cover change (LUCC) has not been considered in long NPP simulation time series. This approach is not rigorous. Because the light use efficiency differs among different land cover types. Third, long time series NDVI data from different sensors have directly been used to simulate the NPP without normalization (Gong et al., 2022).

Considering the above limitations, this study made improvements in the following aspects: (1) Six distinct periods of land cover data were used to simulate the actual NPP (aNPP). (2) NDVI data derived from GIMMS NDVI3g and MODIS NDVI datasets were normalized, and a long time series NDVI dataset from 1982 to 2020 was produced. (3) Three types of NPP were calculated to explore the driving contribution rate to vegetation NPP change. Among, aNPP, reflects the actual carbon accumulation under the prevailing environmental conditions; the potential net primary productivity (pNPP), which represents the maximum productivity merely under optimal climate conditions; and the human-occupied NPP (hNPP), which represents the portion of the NPP influenced by human activities. Notably, hNPP = pNPP – aNPP. To explore the main factors influencing NPP change in China from 1982–2020, we addressed two questions: (1) How did the aNPP change spatiotemporally from 1982–2020? (2) What is the main driving factor underlying spatiotemporally variability at different stages? This study aimed to provide scientific guidance for carbon sink accounting and enhancing the adaptability of vegetation to climate change.

6.2 Data Sources and Methods

6.2.1 Study Area

This study covers a vast area of China (Fig. 6.1), spanning approximately 73.66° to 135.05°E east longitude and 3.86° to 53.56°N. China is located in Northeast Asia and occupies a large territory with a total area of 9.6 million km^2. The regional topography encompasses a gradient from west to east, with the western part dominated by mountains, plateaus, and hills, accounting for 67% of the total land area of China. In contrast, the eastern and southern regions mainly include plains and coastal areas, which account for 33% of the total land area. China has a variety of climate types, including tropical, subtropical, temperate, and plateau climates. The region also exhibits a wide range of land use types, including farmland, forests, grasslands, and construction land. These different land use types directly impact the net primary productivity (NPP) of vegetation. However, in the past decades, climate warming and human interruption, led to land degradation, vegetation NPP and coverage change in some regions of China, for example, 75% of the Qing-Tibet Plateau is an ecologically fragile region with moderate to severe degradation (Yu et al., 2011).

6.2.2 Data Sources

6.2.2.1 NDVI Data

In this study, the data we utilized covered terrestrial China, except for South of China Sea Islands.

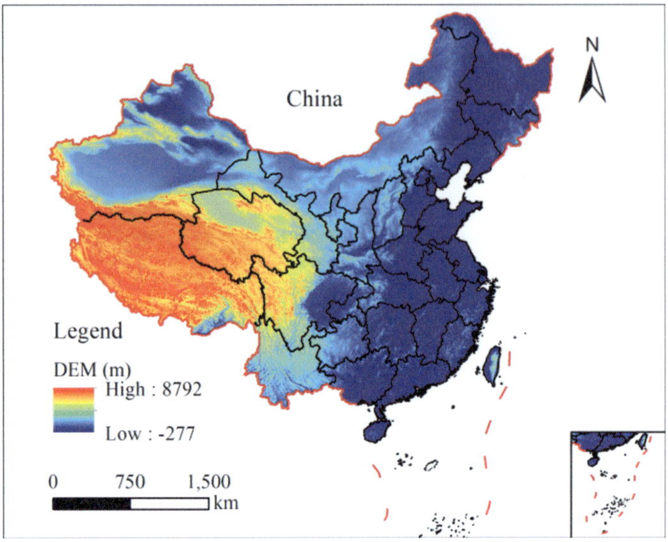

Fig. 6.1 Location of the study area

(1) **NDVI Data Sources and Preprocessing**

The GIMMS NDVI3g dataset, sourced from NASA's Goddard Space Flight Center, provides data at 15-day intervals, with a spatial resolution of 0.083°. This dataset spans from 1981 to 2015. In contrast, the MODIS NDVI data originated from NASA's Earth Observation System EOS/MODIS dataset, which provides data at 16-day intervals. In this study, GIMMS NDVI3g data from 1982 to 2015 and MODIS NDVI data from 2001 to 2020 were used. Both sets of NDVI data were subjected to several preprocessing steps, including data cropping, format conversion, and resampling.

To minimize the effect of cloud cover, we collected two sets of monthly NDVI data using the maximum value composition. The NDVI data is projected as WGS_ 1984 Albers conic projection.

(2) **Consistency Analysis of the GIMMS NDVI3g and MODIS NDVI Datasets**

We chose monthly NDVI data from 2001 to 2015 from the two datasets. In total, 180 months of data were used to conduct consistency analysis. The results showed that both datasets exhibited high consistency for the NDVI peak and trough (Fig. 6.2a). Moreover, there was a significant correlation between the GIMMS NDVI3g and MODIS NDVI datasets (Fig. 6.2b). The regression analysis results were as follows: $NDVI3g = 0.8401\ NDVI_M + 0.0653$ (sample size of 180, $R^2 = 0.9475$, $RMSE = 0.0295$). This linear regression equation was used to interpolate the GIMMS data from 2016 to 2020 using the MODIS NDVI data.

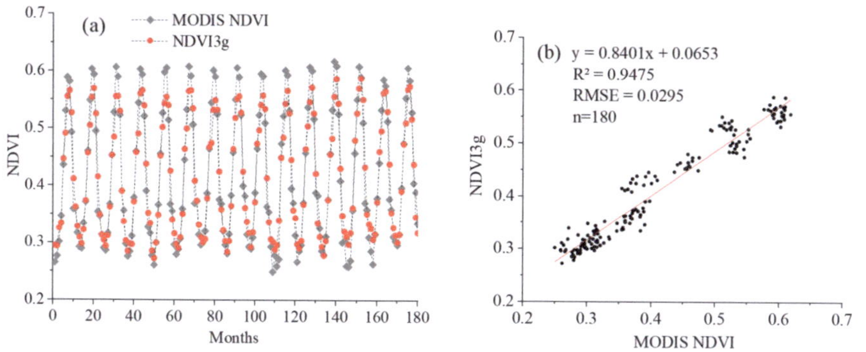

Fig. 6.2 Scatter distribution (**a**) and regression analysis (**b**) of the GIMMS NDVI3g and MODIS NDVI datasets for China

6.2.2.2 Meteorological Data

Meteorological data were mainly obtained from the China Meteorological Data Sharing Service System (http://cdc.cma.gov.cn). The data included the monthly average temperatures, total monthly precipitation data, and monthly total solar radiation measurements from 1982 to 2020. The data were processed through interpolation combined with digital elevation model (DEM) data using the AUSPLINE interpolation program as it was suitable for interpolation in the complex region of China (Huang et al., 2016).

6.2.2.3 Land Use/Land Cover Change Data

The LUCC data were derived from the Data Center for Resources and Environmental Sciences, Chinese Academy of Sciences (http://www.resdc.cn/). The dataset covers six periods, namely, 1990, 1995, 2000, 2005, 2010, and 2015, and the spatial resolution was 1 km. The LUCC data comprise six categories: cultivated land, forestland, grassland, water bodies, urban and built-up land, and unused land. In this study, deserts and water bodies were not considered.

LUCC data correspond to different periods. For example, the NPP simulation from 1982 to 1992 used the LUCC data from 1990, while the NPP simulation from 1993 to 1997 used the LUCC data from 1995, and so on. This time-matching approach reduced the influence of land use and cover change on the NPP simulations in long time series.

6.2.2.4 Socioeconomic Data

Socioeconomic data, including population density (POP) and gross domestic product (GDP) data, were derived from China's National Bureau of Statistics (http://www.stats.gov.cn/).

6.2.3 *Vegetation NPP Simulation Model*

6.2.3.1 ANPP Simulation

In this study, the CASA model, which is widely applied in regional and global terrestrial carbon cycle studies, was used to simulate the aNPP. The CASA model is based on the principle of plant photosynthesis and considers various factors, such as the absorbed photosynthetically active radiation (APAR), temperature stress coefficient, water stress coefficient, and maximum light use efficiency (ε), to calculate the net primary productivity (NPP) of vegetation. The equations are as follows:

$$\text{ANPP}\left(\text{g C/m}^2/\text{year}\right) = \text{APAR} \times \varepsilon \tag{6.1}$$

$$\text{APAR}\ \left(\text{MJ/m}^2\right) = \text{SOL} \times \text{FPAR} \tag{6.2}$$

where APAR is the photosynthetically active radiation, SOL is the total solar radiation, and FPAR is the proportion of the photosynthetically active radiation absorbed by vegetation, which is usually set to 0.5, indicating that vegetation absorbs half of the total radiation. The light use efficiency (ε) is one of the key parameters used to simulate the NPP. For more specific equations, refer to (Zhu et al., 2006). The maximum light uses efficiency (ε_{max}) values for deciduous coniferous forests, evergreen coniferous forests, shrublands are 0.485 g C/MJ, 0.389 g C/MJ, 0.429 g C/MJ, and 0.542 g C/MJ for grassland, farmlands, unused land, building land and water.

6.2.3.2 Potential NPP Simulation (pNPP)

The pNPP is a measure of the maximum production capacity of vegetation under the influence of climatic factors alone, excluding the effects of human activities. The pNPP can be simulated using a climate productivity model. In previous research, the pNPP has been utilized to quantitatively assess land degradation, yielding credible evaluation outcomes (Zhou et al., 2017). The Miami model, which originates from statistical correlations between observed NPP data and temperature/precipitation data, is a widely employed pioneering model. Over the last three decades, this model has yielded a fundamental dataset, providing reasonable estimates of global productivity patterns (Adams et al., 2004). Therefore, in this study, the pNPP was simulated by the Miami model.

6.2.3.3 Human-Occupied NPP (hNPP)

In this study, the hNPP was defined as the difference between the pNPP and aNPP, reflecting the impact of human activities on the vegetation productivity. A positive difference indicates that human activities impede the actual productivity from reaching its potential, suggesting that human activities are detrimental. Conversely, a negative difference suggests that human activities enhance the actual productivity beyond the potential capacity, indicating a beneficial influence. The residual method was used to calculate the hNPP (Xu et al., 2010; Zhou et al., 2015), which can be expressed as follows:

$$\text{hNPP} = \text{pNPP} - \text{aNPP} \tag{6.3}$$

For hNPP > 0, human activities are harmful to the aNPP and prevent increases, and vice versa.

6.2.3.4 Verification of the aNPP Simulation Accuracy

The field survey NPP was used to validate the accuracy of the aNPP simulation results. On the one hand, the field survey data originated from our sampling from late July or early August from 2017 to 2020. On the other hand, survey data for the 2005 to 2016 period were obtained from the Ecological Environment Monitoring Center. The recorded data for each plot included the latitude, longitude, elevation, land cover type, coverage, and fresh weight according to a previous field survey. The sampling sites included alpine grassland, alpine meadow, temperate grassland, temperate desert, forest, and farmland sites. The dry weight was determined after drying the sample to a constant weight at 65 °C for 48 h in the laboratory. According to previous studies, the carbon conversion rate was set to 0.45 (Li et al., 2016). Then, the measured NPP (g C/m^2) was obtained. Finally, 324 field-measured aNPP data points were used to validate the model. To assess the accuracy of the model predictions, the RMSE and R^2 were used.

6.2.4 Contribution Rate Calculate of Climate and Human Acclivities on aNPP Change

The driving contribution rates of climate and human activities to grassland aNPP were quantitatively separated. The specific method is shown in Table 6.1, referring to our previous study (Zhou et al., 2017).

In this study, aNPP, pNPP, and hNPP were used to quantitatively evaluate the driving contribution rates of climate and human activities to vegetation aNPP changes. A positive trend in pNPP indicates a positive impact of climate change on the enhancement of aNPP, and vice versa. Similarly, a positive trend in hNPP implies that human activities are contributing to a decrease in aNPP, and vice versa. Six scenarios were used to calculate the contribution of climate change and human activities to aNPP variation.

6.2.5 Relative Importance of Influence Factors to Vegetation aNPP Change

Random forest (RF) is an ensemble learning method that enhances the accuracy and stability of predictions by constructing multiple decision trees. When handling ecological environment data, RF can effectively address non-linear relationships, missing values, and high-dimensional data issues. Meanwhile, the feature importance assessment function of RF can typically be conducted based on methods such as mean reduced impurity (MDI) or Ranking Importance. Therefore, RF can help identify feature factors that have a significant impact on NPP.

In this study, we utilized the feature importance assessment function of RF to quantitatively calculate the importance of driving factors.

6.3 Results

6.3.1 aNPP Simulation Accuracy Analysis

To assess the estimation accuracy of NPP, field-measured NPP data were employed for comparative analysis. The results indicated satisfactory correspondence between the model-estimated values and the measured values (Fig. 6.3). The coefficient of determination (R^2) was 0.7629, which indicates statistical significance ($P < 0.01$). These findings indicated the reliability of the aNPP simulations using the CASA model in China.

Table 6.1 Scenario settings for the relative contributions of climate and human activities to grassland aNPP change

	Scene	S_{pNPP}	S_{hNPP}	Climate contribution (%)	Human contribution (%)	Scene description
Increase in aNPP	1	> 0	> 0	100	0	Climate change dominated NPP increase
	2	> 0	< 0	$\frac{\lvert\Delta pNPP\rvert}{\lvert\Delta pNPP\rvert+\lvert\Delta hNPP\rvert} \times 100\%$	$\frac{\lvert\Delta hNPP\rvert}{\lvert\Delta pNPP\rvert+\lvert\Delta hNPP\rvert} \times 100\%$	Combined contribution
	3	< 0	< 0	0	100	Human activities dominated NPP increase
Decrease in aNPP	4	< 0	< 0	100	0	Climate change dominated NPP decrease
	5	< 0	> 0	$\frac{\lvert\Delta pNPP\rvert}{\lvert\Delta pNPP\rvert+\lvert\Delta hNPP\rvert} \times 100\%$	$\frac{\lvert\Delta hNPP\rvert}{\lvert\Delta pNPP\rvert+\lvert\Delta hNPP\rvert} \times 100\%$	Combined contribution
	6	> 0	> 0	0	100	Human activities dominated NPP decrease

Note aNPP is actual NPP, pNPP is potential NPP, hNPP is human-appropriated NPP. Scenarios 1–3 represent the actual NPP increase, where ΔpNPP and ΔhNPP represent the variation in potential NPP and human-appropriated NPP, respectively, in a certain period. The units of aNPP, pNPP, and hNPP are g C/m^2

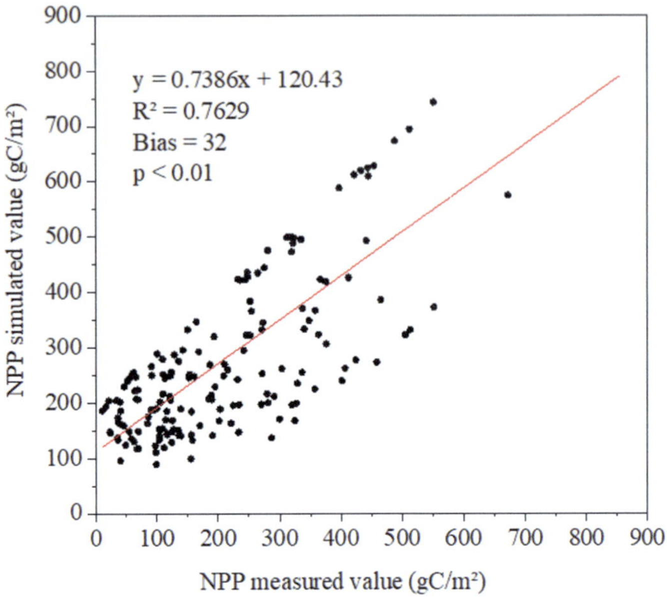

Fig. 6.3 Regression relationship between the measured and model-simulated NPP values

6.3.2 Spatiotemporal Changes of the Vegetation aNPP in China

From 1982 to 2000, the mean aNPP in China was approximately 501.0 g C/m^2, and the annual total aNPP reached 378.2 Tg C. However, from 2001 to 2020, the aNPP slightly increased, reaching 517.1 g C/m^2, while the annual total aNPP reached 391.1 Tg C.

The spatial distribution of the aNPP in China varied (Fig. 6.4a, b). Overall, there was a decreasing trend in the annual mean aNPP from southeast to northwest, similar to previous studies. Regions with high aNPP values exceeding 1000 g C/m^2 were mainly concentrated in southern and southwestern China. aNPP values lower than 50 g C/m^2 were mainly distributed in the northwestern and western parts of the Qinghai–Tibet Plateau.

From 1982 to 2000, the aNPP exhibited an extremely significant increase ($P < 0.01$) and showed a significant increase from 2001 to 2020 ($P < 0.05$) (Fig. 6.4c).

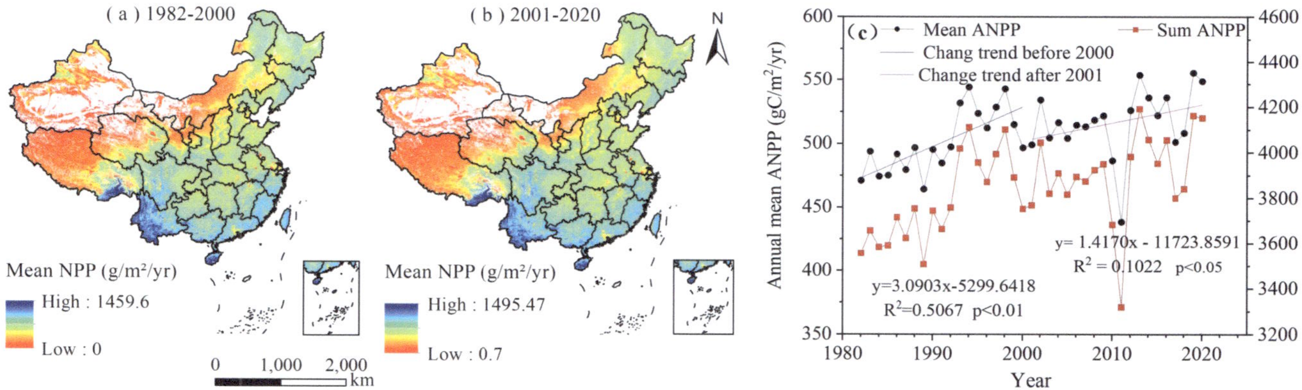

Fig. 6.4 Spatial distribution of vegetation aNPP in China **a** from 1982 to 2000, **b** from 2001 to 2020, and **c** changes in the annual mean aNPP and annual total aNPP

6.3.3 Interannual Variation Trend of the Three NPPs in China

6.3.3.1 Change Trends of NPPs from 1982 to 2000

From 1982 to 2000, the aNPP showed an increasing trend (Fig. 6.5a; Table 6.2), and the increase rate reached to 83.6%. In terms of the spatial distribution, there was a decreasing gradient from southeast to northwest. Notably, southern China exhibited the greatest increase in the aNPP (10–120 g C/m^2).

The pNPP showed an increasing trend (Fig. 6.5b; Table 6.2), with the increase rate reaching to 63.3%. In terms of the spatial distribution of the pNPP changes, there was a notable increase (3–14 g C/m^2) in southwestern Tibet, southern China. The regions with decreased pNPP values were mainly distributed in Central China.

The hNPP mainly showed a declining trend (Fig. 6.5c; Table 6.2), and the decrease rate reached to 77.7%, widely distributed across China. Regions with a declining trend occurred in northern Tibet.

Compared with the changes in the aNPP, pNPP, and hNPP, both climate change and human activities caused an increase in the aNPP.

6.3.3.2 Change Trends of NPPs from 2001 to 2020

From 2001 to 2020, the aNPP showed an increasing trend (Fig. 6.5a), and the increase rate reached to 63.96% (Table 6.3). The Yellow River Basin exhibited a significant increase in the aNPP (10–120 g C/m^2). In contrast, the aNPP in most of regions, including the southeastern hills, North China Plain, Shandong hills, Changbai Mountains, Hengduan Mountains, southeastern Qinghai–Tibet Plateau, western Sichuan, and surrounding areas of the Tianshan Mountains, exhibited a decreasing trend (− 10 to − 1 g C/m^2).

The pNPP showed an increase trend, increase rate reached to 90.17% (Fig. 6.5b; Table 6.3). The largest increase in the pNPP (3–14 g C/m^2) was recorded in the Sichuan Basin, South China, and Northeast China. In contrast, the pNPP decreased (− 3 to − 1 g C/m^2) in southwestern Yunnan, southeastern Tibet, northern Henan, and northeastern Shandong.

There was an overall growth trend (63.16%) in the hNPP across China (Fig. 6.5c; Table 6.3). The central region exhibited relatively small increases (0–1 g C/m^2).

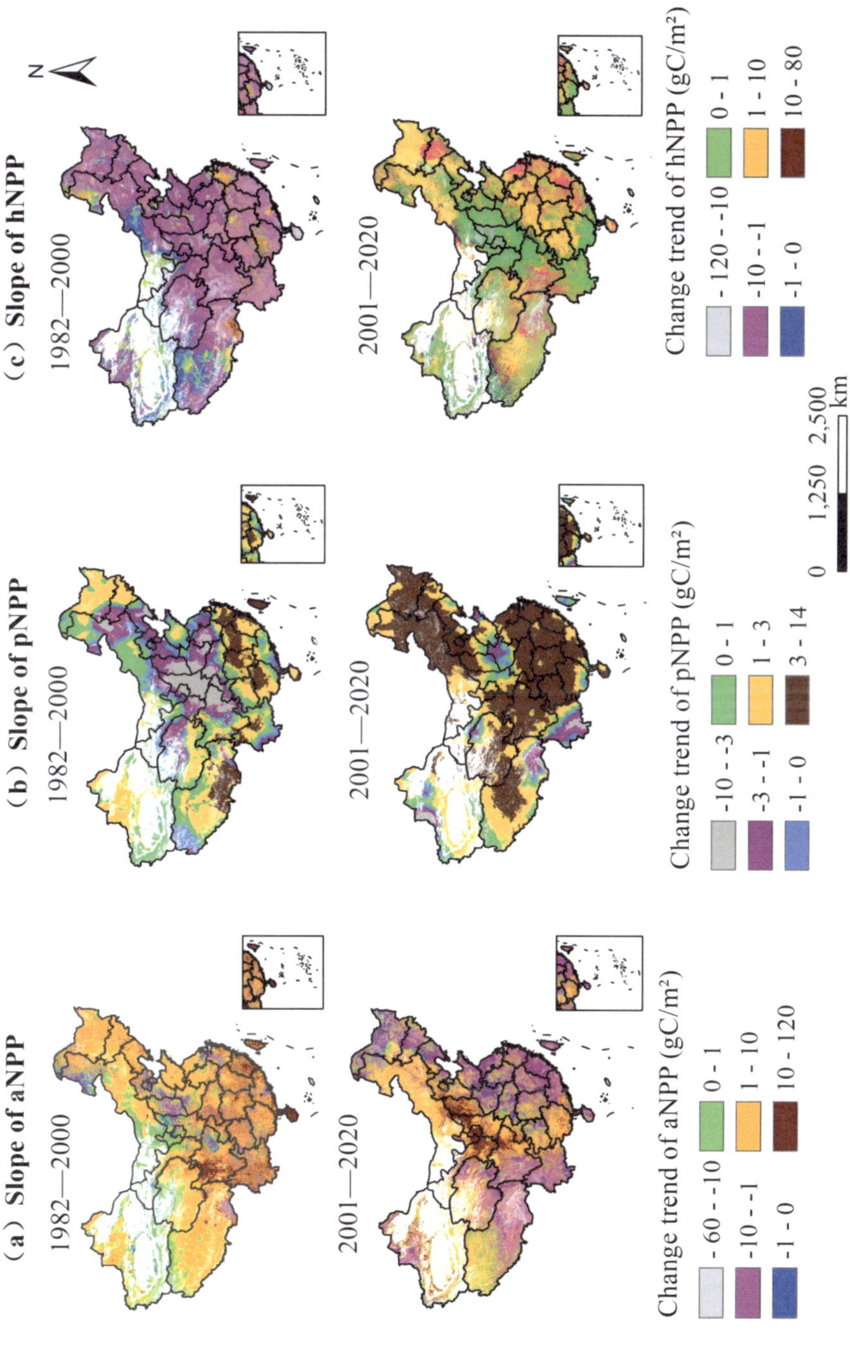

Fig. 6.5 Spatial distributions and change trends of the aNPP (**a**), pNPP (**b**), and hNPP (**c**)

Table 6.2 Statistics of the interannual trends of the three NPP categories in China from 1982 to 2000

Change trend of the three NPP categories (g C/m^2) (1982–2000)	Change in the aNPP (%)	Change in the pNPP (%)	Change in the hNPP (%)
− 60 to − 10	1.5	9.6	8.2
− 10 to − 1	9.4	14.1	56.9
− 1 to 0	5.5	13	12.6
0–1	14.1	22.2	8.4
1–10	60.9	31.3	12.1
10–120	8.6	9.8	1.7

Table 6.3 Statistics of the interannual trends of the three NPPs in China from 2001 to 2020

Change trend of the three NPPs (g C/m^2) (2001–2020)	Change in the aNPP (%)	Change in the pNPP (%)	Change in the hNPP (%)
− 60 to − 10	4.98	1.5	4.61
− 10 to − 1	24.12	3.83	24.98
− 1 to 0	6.94	4.5	7.25
0–1	8.79	8.28	8.12
1–10	46.71	27.17	45.99
10–120	8.46	54.72	9.05

6.3.4 *Contributions of Climate and Human Activities to Changes in the Actual NPP*

6.3.4.1 Quantitative Analysis of the Contribution Rate from 1982 to 2000

From 1982 to 2000, the increase in the aNPP was influenced by various factors (Fig. 6.6a; Table 6.4), including climatic factors (contribution was 12.24%), human factors (34.26%), and climatic–human factors (19.79–33.71%). Climatic factors played a significant role in promoting aNPP growth in most parts of Xinjiang and western Inner Mongolia. The increase in the aNPP in South China, Shandong, Heilongjiang, and the Yunnan-Guizhou Plateau was influenced by climatic–human factors.

The decrease in the aNPP (Fig. 6.6b; Table 6.3) between 1982 and 2000 was mainly caused by human activities (contribution was 50.43%) and climatic factors (26.70%). In Central China, the decrease in the aNPP was mainly due to climatic factors, while the decline in the rest of the country was mainly the result of human activities such as urbanization, deforestation, and overgrazing.

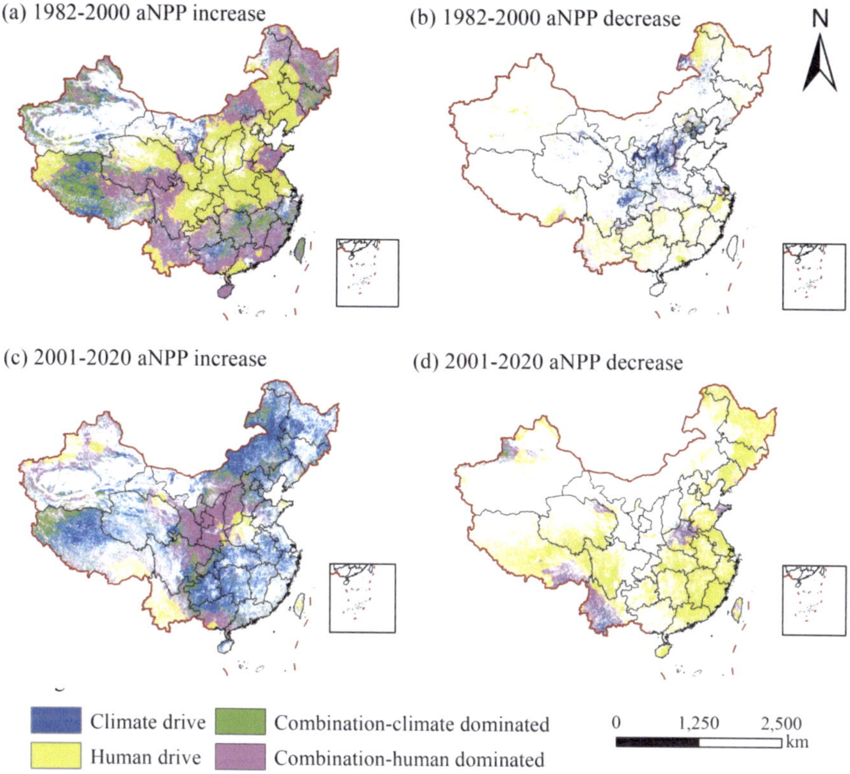

Fig. 6.6 Contributions of climate and human factors to NPP increase and decrease in China from 1982 to 2020

Table 6.4 Contribution rates of climate and human activities to the actual NPP in China

Period (year)	aNPP change	Contribution rate (%)		
		Climatic change	Human activity	Combined (climate-human)
1982–2000	Increase (Tg C)	12.24	34.26	19.79–33.71
	Decrease (Tg C)	26.70	50.43	8.1–14.77
2001–2020	Increase (Tg C)	43.90	6.21	22.24–27.65
	Decrease (Tg C)	2.66	83.75	2.02–11.57

6.3.4.2 Quantitative Analysis of the Contribution Rate from 2001 to 2020

From 2001 to 2020, climatic factors were the main drivers of the increase in the aNPP (contribution was 43.90%), particularly in Xinjiang, Inner Mongolia, and the Qinghai–Tibet Plateau (Fig. 6.6c; Table 6.4). Human activity factors also played a role in certain areas (6.21%), particularly in Yunnan, Henan, southeastern Tibet, and northwestern Xinjiang (Fig. 6.6c; Table 6.4).

Human activities were identified as the primary cause of the decrease in the aNPP, contribution reached to 83.75% of the overall decrease (Fig. 6.6d; Table 6.4). In certain areas, such as southwestern Yunnan, southeastern Tibet, northeastern Shandong, and most of Henan, the decline was influenced by the interplay between climatic and human factors, which ranged from 2.02 to 11.57% (Fig. 6.6d; Table 6.4). These findings underscore the negative impacts of human activities on the vegetation productivity.

6.3.5 *The Quantitative Assessment of Grassland NPP Change in Tibetan for the Regions that Always Maintained Grassland*

6.3.5.1 Quantitative Assessment of aNPP Change from 1982 to 2000

From 1982 to 2000, the area that always remained as grassland was 1,289,700 km^2, and the proportion of area with increasing and decreasing aNPP was 97.4% and 2.6%, resulting in an increase and decrease of 58.9 Tg C and 0.4 Tg C, respectively, and a net increase of 58.5 Tg C in total aNPP.

For the region with aNPP increase, the contribution of climate and human activities was 7.3% and 13.0%, respectively, and the contribution rates of the two interacting together were 28.4% and 51.3%, respectively. The dominant climatic influence was observed mainly in the central Tibetan Plateau, while human activities were concentrated in the northern region of Qinghai Province and the western region of Tibet (Fig. 6.7a).

For the region with aNPP decrease, climate contributed 4.5% to the decline in aNPP, while human activities accounted for 80%. In addition, the contribution rates of the interactive influence together were 3.7% and 11.9%, respectively. The regions with pronounced human activities influence were in the southern and northwestern parts of the Tibetan Plateau (Fig. 6.7d).

In general, human activities emerged as the predominant factor causing increases (64.3%) and decreases (91.9%) in grassland aNPP.

6.3.5.2 Quantitative Assessment of the aNPP Change from 2001 to 2020

During 2001–2020, the area that always remained as grassland was 1,506,600 km^2, and the area proportions with increasing and decreasing aNPP were 65.7% and 34.3%, resulting in an increase and decrease of 62.7 Tg C and 48 Tg C, respectively, and a net increase of 14.7 Tg C in the total aNPP.

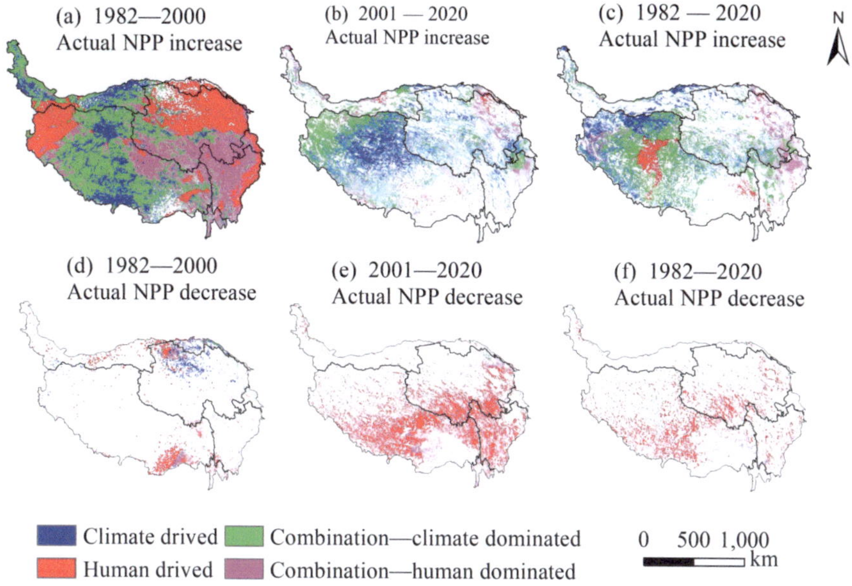

Fig. 6.7 Driving contributions of climate and human activities to aNPP change on the Qinghai-Tibet Plateau

For the region with an increasing grassland aNPP trend, the proportional contributions of climate and human activities accounted for 22.3% and 3.8%, respectively, and the contribution rates of the two interacting together were 30.4% and 43.5%, respectively. Therefore, climate-dominated grassland aNPP increased and was widely distributed, and the areas dominated by human activities were scattered in northeastern Qinghai and southern Tibet (Fig. 6.7b).

For the regions with a decreasing grassland aNPP trend, the respective contributions of climate and human activities were quantified as 0.02% and 86.8%, respectively, while their combined interactions were 0.18% and 13.0%, respectively. Therefore, human activities dominated the aNPP decrease and were widely distributed (Fig. 6.7e).

Overall, climate dominated the aNPP increase (contribution rate was 52.7%), and human activities dominated the aNPP decrease (contribution rate was 99.8%).

6.3.5.3 Quantitative Assessment of the aNPP Change from 1982 to 2020

From 1982 to 2020, the area that always remained as grassland was 1,287,400 km^2, and the proportion of area with increasing and decreasing aNPP was 79.3% and 20.7%, resulting in an increase and decrease of 82.2 Tg C and 13.5 Tg C, respectively, and a net increase of 68.8 Tg C.

For the region with an increasing grassland aNPP trend, the contribution rates of climate and human activities were 16.8% and 6.5%, respectively, and the contribution rates of the two interacting together were 33.3% and 43.4%, respectively (Fig. 6.7c). Therefore, their respective contributions of climate and human activities being proportionally equal (50.1% vs. 49.9%).

For the region with the decreasing grassland aNPP trend, the contribution rates of climate and human activities were 0.2% and 82.3%, respectively, while their combined interaction contributed 0.5% and 17.0%, respectively (Fig. 6.7f).

In general, human activities dominated the aNPP decrease, with a contribution rate of 99.3%.

6.3.6 Comprehensive Evaluation of the Contribution Rate to aNPP Variation During Three Periods

From 1982 to 2000, for the all-region of grassland, aNPP dramatically increased by 70.58 Tg C, among which climate, human activities (except LUCC) and LUCC led to aNPP increases of 25.65 Tg C, 37.60 Tg C, and 7.33 Tg C, respectively. The contribution rates of these factors were 36.3%, 53.3% and 10.4%, respectively.

From 2001 to 2020, aNPP increased by 14.55 Tg C, among which climate and LUCC led to aNPP increases of 32.63 Tg C and 0.09 Tg C, and the contribution was 224.2% and 0.6%, respectively. However, human activities (except LUCC) led to a decrease in aNPP of 18.16 Tg C, and its contribution rate was − 124.8%.

From 1982 to 2020, grassland aNPP dramatically increased by 80.25 Tg C, among which climate and LUCC led to aNPP increases of 65.92 Tg C and 27.62 Tg C, and the contribution was 82.2% and 34.4%, respectively. However, LUCC led to aNPP decrease of 13.29 Tg C, and its contribution rate was − 16.5% (Fig. 6.8).

6.4 Discussion

6.4.1 Impact of Climatic Factors on the Actual NPP

Climate change is the key factors impacting the distribution and dynamics of vegetation (Song & Ma, 2011). Table 6.5 reveals that 20.6% of China showed a negative correlation between aNPP and precipitation, possibly due to intense precipitation causing nutrient loss and soil erosion (Xu et al., 2016), consequently impacting plant growth. In contrast, 79.4% showed a positive correlation, among, 35.09% showed a significant positive correlation, mainly distributed in the northwestern of China (Fig. 6.9a), because the dry climate inhibiting vegetation growth, increase of precipitation can improve soil moisture and promote vegetation growth.

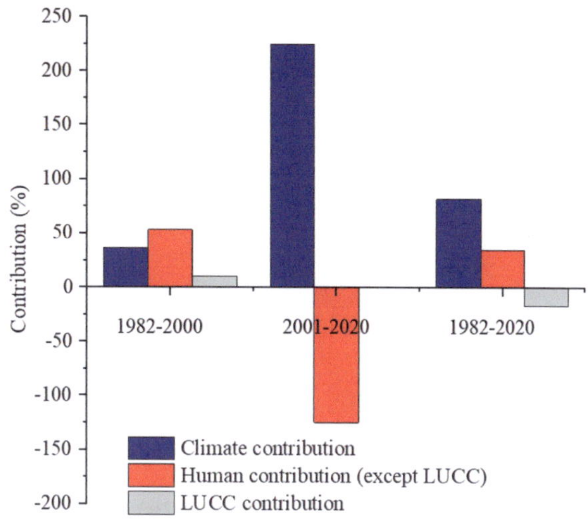

Fig. 6.8 Statistical chart of the driving contribution rates of grassland aNPP on the Tibetan Plateau

Table 6.5 Correlation coefficients between the aNPP and precipitation and temperature

Correlation with precipitation	Proportion (%)	*P* value	Correlation with temperature	Proportion (%)	*P* value
− 0.76 to − 0.403	0.45	< 0.01	− 0.95 to − 0.403	3.17	< 0.01
− 0.403 to − 0.312	1	< 0.05	− 0.403 to − 0.312	2.39	< 0.05
− 0.312 to 0	19.15	> 0.05	− 0.312 to 0	18.23	> 0.05
0–0.312	44.31	> 0.05	0–0.312	35.69	> 0.05
0.312–0.403	10.87	< 0.05	0.312–0.403	12.7	< 0.05
0.403–0.97	24.22	< 0.01	0.403–0.94	27.81	< 0.01

Regarding the temperature (Table 6.5), 23.79% of the regions, such as Sichuan, Yunnan, and Southeast China, showed a negative correlation between the aNPP and temperature (Fig. 6.9b). Because, elevated temperatures inhibit photosynthesis, nitrogen consumption, and the peroxide quota, leading to decreased plant productivity (Xu & Zhou, 2006), for these regions with original higher temperature. Conversely, 76.21% showed a positive correlation, mainly distributed in the Qinghai-Tibet Plateau, northern and northeastern of China (Fig. 6.9b). For these cold regions, the increasing temperatures influenced hydrological and ecological processes by promoting soil freeze–thaw cycles, thus enhancing plant growth conditions and nutrient cycling and yielding an increased NPP.

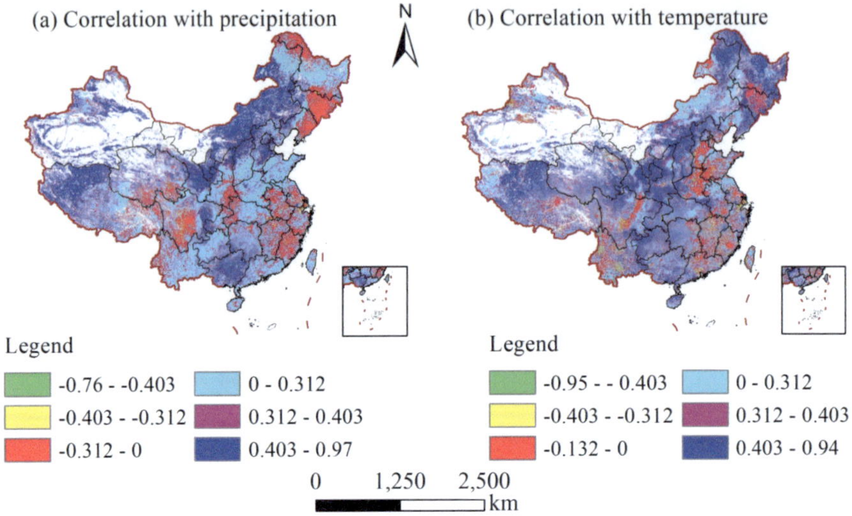

Fig. 6.9 Spatial distributions of the correlation between the aNPP and **a** precipitation, and **b** temperature. *Note* A correlation coefficient between 0.312 and 0.403 indicates a significant correlation ($p <$ 0.05), and a coefficient higher than 0.403 indicates an extremely significant correlation ($p < 0.01$)

6.4.2 Impact of Human Activities on the Actual NPP

Human activities are also significant factors influencing the recovery or degradation in vegetation (Naeem et al., 2020). About 59.56% of China, there was a negative correlation between the population density and NPP (Fig. 6.8a; Table 6.6), mainly distributed in the Qinghai–Tibet Plateau, Sichuan, Inner Mongolia, and Southeast China. Because the high population density leads to excessive land development, overgrazing, and deforestation, causing land degradation and NPP decrease (Li et al., 2016). About 40.44% of the regions showed a positive correlation between the population density and NPP (Fig. 6.10a; Table 6.6), notably in the North China Plain and the middle and lower reaches of the Yangtze River Plain. Because human agricultural activities contribute positively to the NPP increase in these regions (Ge et al., 2021).

Concerning the relationship between the GDP and NPP (Fig. 6.10b; Table 6.6), approximately 61.47% of the regions showed a positive correlation, which was similar to previous research (Jiang et al., 2013). The regions included the Yellow River Basin and Central China, the positive correlation may be influenced by ecological restoration projects such as the Grain to Green Project (GTGP) and Grazing to Grassland Project (GWP), which aim to effectively protect and restore vegetation. However, in economically developed regions such as Henan, Anhui, Jiangsu, and Hubei, high urbanization levels contribute to negative correlations between the GDP and NPP.

Table 6.6 Correlation coefficients between the aNPP and population density and GDP

Correlation with population	Proportion (%)	*P* value (population)	Correlation with GDP	Proportion (%)	*P* value (GDP)
– 1 to – 0.875	4.49	$P < 0.01$	– 1 to – 0.875	0.68	$P < 0.01$
– 0.875 to – 0.755	6.09	$P < 0.05$	– 0.875 to – 0.755	1.75	$P < 0.05$
– 0.755 to 0	48.98	$p > 0.05$	– 0.755 to 0	36.81	$p > 0.05$
0–0.755	40.44	$p > 0.05$	0–0.755	60.76	$p > 0.05$
0.755–0.875	3.70	$P < 0.05$	0.755–0.875	10.15	$P < 0.05$
0.875–1	1.92	$P < 0.01$	0.875–1	6.52	$P < 0.01$

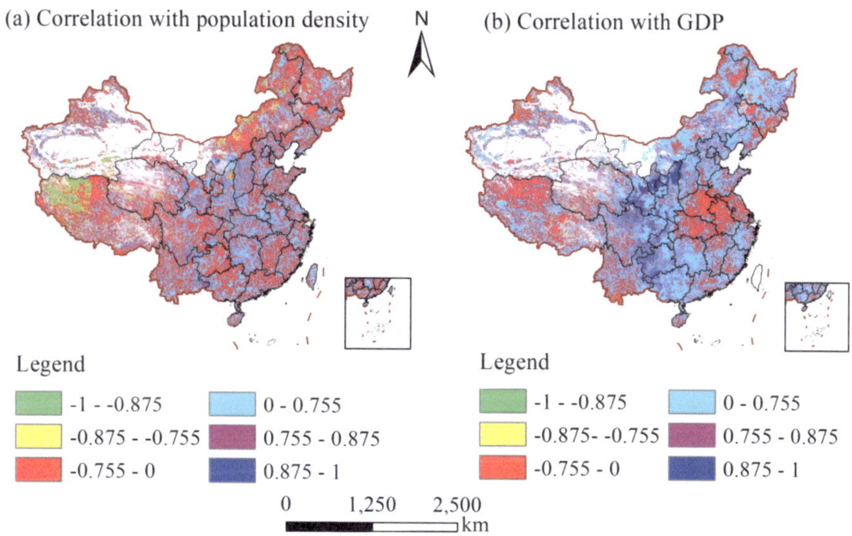

Fig. 6.10 Spatial distributions of the correlation between aNPP and **a** POP and **b** GDP. *Note* A correlation coefficient between 0.755 and 0.875 indicates a significant positive correlation ($p < 0.05$), and a correlation coefficient higher than 0.875 indicates an extremely significant positive correlation ($p < 0.01$)

6.4.3 Importance of the Influencing Factors for aNPP Change

The factors importance for aNPP change was calculated based on RF during the 2001–2020 period. Results showed that natural factors (50.9%) and human activities (49.1%) contributed almost equally to the aNPP variation, which was consistent to our previous research in Qinghai-Tibet Plateau (Zhou et al., 2024). Among them, climatic factors and socioeconomic factors exhibited the highest contributions, reaching to 36.3% and 31.3%, respectively.

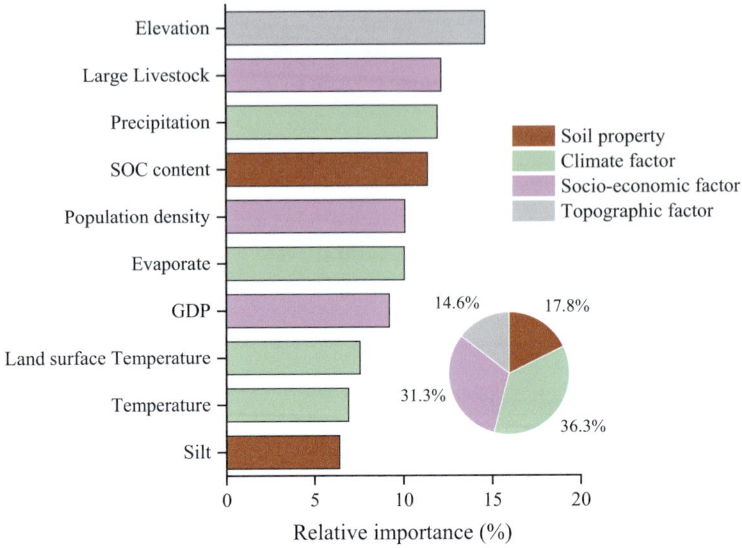

Fig. 6.11 Relative importance of the influencing factors according to the RF model

Among the climatic factors (Fig. 6.11), temperature influences plant growth periods and photosynthetic rates, its contribution reached to 6.88%. The contribution of the land surface temperature (LST) and precipitation reached to 7.53% and 11.89%, because LST are directly related to soil moisture, water supply, and vegetation water requirements (Dai et al., 2022). In arid or semiarid regions, intense evaporation may lead to water stress, thereby reducing the NPP, its contribution was 10.02%.

For the socioeconomic factors, the contribution of GDP was 9.17%, because of the increase in the GDP can accelerate agricultural, industrial, and urbanization processes, altering land cover types and affecting the NPP (Imhoff et al., 2004). For the population density, its importance was 10.06%, the increased demand for land resources in densely populated areas may disrupt natural vegetation, leading to a reduced NPP in those regions (Wu & Wu, 2023). The importance of livestock industry was 12.11%, the development of livestock industry could aggravate the grazing pressure, potentially causing grassland degradation and lowering the NPP of grassland ecosystems (Lyu et al., 2022).

Overall, the altitude was the single most influential factor of the NPP, accounting for 14.58%. Figure 6.5a also demonstrates that the NPP was generally lower in high-altitude areas, while it may peak within moderate altitude ranges due to the favorable temperature conditions. As a crucial geographical gradient, the altitude profoundly impacts climate, soil, vegetation distribution, and productivity (Mayor et al., 2017), thereby influencing the overall pattern of the NPP. With increasing altitude, temperatures decrease, directly affecting plant growth and photosynthesis. Moreover, climate and vegetation type variations at different altitudes lead to corresponding changes in the NPP. Differences in the soil texture, thickness, and nutrient content between high- and low-altitude areas also impact the NPP.

Furthermore, soil properties play a significant role that cannot be overlooked, the importance of SOC and silt were 11.5%, 6.5%. The soil organic carbon content and soil texture represent nutrient levels (Ge et al., 2019) and water retention capacity (Rawls et al., 2003), respectively, which exert crucial impacts on plant growth and photosynthetic rates.

6.4.4 *Strategies to Increase Vegetation Productivity*

The primary human activities affecting alpine grassland net primary productivity (NPP) on the Tibetan Plateau are grazing and human-induced land use and land cover change (LUCC) (Chen et al., 2013). For instance, since 1978, livestock production has increased by nearly 300%(Du et al., 2004). This surge in human and livestock populations has led to substantial ecological changes, including overgrazing, fertilization for enhanced forage production (Niu et al., 2009), rangeland privatization (Yan & Wu, 2005), and the settlement of previously nomadic communities. These developments have collectively intensified the human impact on the plateau's ecosystems. For example, severe overgrazing is considered a major driver of grassland degradation and even desertification (Song et al., 2009). Additionally, since the 1960s, extensive commercial logging has significantly reduced forest cover in the eastern parts of the Tibetan Plateau (Wu & Liu, 1998).

Increasing population has accelerated urbanization, particularly in Tibet, where the urban population rose from 0.4 million in 1995 to 1.0 million in 2006 (Fan et al., 2010). Meanwhile, the warming trend intensified between 1982–2000 and 2001–2020, with the temperature increase rate rising from 0.48 °C/decade to 1.0 °C/decade, and the precipitation increase rate climbing from 1.3 mm/yr to 3.2 mm/yr. As a result, the contribution rate of climate-dominated aNPP increase jumped from 36.4% to 224.3%, largely because the current growing-season temperature on the Tibetan Plateau remains below the optimal temperature for vegetation growth (Huang et al., 2019).

In the context of ongoing global warming, key questions arise: Will the growing-season temperature on the Tibetan Plateau remain below the optimum for vegetation growth? Can the climate-driven enhancement of vegetation NPP be sustained, and if so, for how long? The answers depend not only on the pace of warming but also on the adaptability of vegetation to changing climatic conditions.

To mitigate human-induced ecosystem degradation on the Tibetan Plateau, China has implemented several large-scale conservation initiatives, including the Natural Forest Conservation Program (NFCP) launched in 1998, the Grain for Green Program (GTGP) in 1999, and the Returning Grazing Land to Grassland Program (RGGP) in 2003. Research indicates that land cover modifications resulting from these programs, particularly afforestation and reforestation efforts, have significantly enhanced carbon sequestration capacity (Liu et al., 2008). This effect is especially pronounced in the eastern Himalayas and Hengduan Mountains regions, where forest coverage is predominant.

Furthermore, studies demonstrate that grassland restoration effectively improves soil water conservation (Ouyang et al., 2016). This conclusion is substantiated by long-term fencing experiments, which revealed that soil moisture content in enclosed, non-degraded grasslands was 18.3 to 27.8% higher than in severely degraded grasslands (Xu et al., 2013). Additional research has documented evident grassland recovery in Maqu County following RGGP implementation (Wang et al., 2009).

In this study, the actual net primary production (aNPP) on the Tibetan Plateau increased by 70.58 Tg C during 1982–2000, with human activities (excluding LUCC) and LUCC collectively accounting for 63.7% of this increase. In the subsequent period (2001–2020), the aNPP rise was considerably smaller, at only 14.6 Tg C, and the contribution of human activities shifted to negative (− 124.4%). A comparison of the two periods reveals a significant slowdown in aNPP growth. This finding is consistent with previous research (Chen et al., 2020), which reported that the proportion of human-dominated NPP reduction was twice that of human-dominated NPP increase, alongside an expansion in the affected area. These results underscore the critical importance of implementing precise and stringent grassland management strategies for the Tibetan Plateau amid ongoing climate warming and intensifying human pressures.

First, overgrazing should be mitigated to achieve a sustainable balance between grassland carrying capacity and livestock density. Second, pastoral livelihoods should be diversified to facilitate a steady increase in herders' income. Third, ecological compensation standards should be determined based on the effectiveness of grassland restoration in each pasture, so as to incentivize herders' participation in degradation control initiatives. Fourth, rotational grazing, stall feeding, and cultivated grassland establishment should be promoted according to local hydrothermal conditions, along with strengthened management of rodent and insect pests. The selection of grass species for planting must be tailored to local conditions, considering that alpine meadows are more sensitive to temperature warming (Chang et al., 2012), while alpine steppes respond more strongly to precipitation variability (Peng et al., 2009). Furthermore, ecological awareness among herders should be enhanced through targeted publicity and education on grassland conservation. The implementation of these coordinated strategies and supportive government policies will contribute to the sustainable development of grassland resources and the maintenance of ecological security on the Tibetan Plateau.

One limitation of this study is the low number of meteorological stations in northwestern Tibet, which increases the uncertainty of the spatial distribution of climate variables, although DEM data were used as an auxiliary variable to drive spatial interpolation in this study. Therefore, in future studies, data fusion based on meteorological station observation data, interpolation surface data, or reanalysis data will be used to decrease the uncertainty.

6.5 Conclusions

This study, based on three NPP categories, provided a method to quantitatively calculate the contribution rates of climate and human activities to aNPP changes from 1982 to 2020 in China. The results showed that: (1) from 1982 to 2000, the annual mean aNPP was 501.0 g C/m^2, and the annual total aNPP was 378.2 Tg C. From 2001 to 2020, the annual mean aNPP was 517.1 g C/m^2, and the annual total aNPP was 391.1 Tg C. (2) Compared the two-period stage, the contribution of the climate to the increase of the aNPP significantly increased from 12.24 to 43.90%, while the contribution of human activities to the decrease of the aNPP clearly increased from 34.26% to 83.75%. (3) The aNPP was significantly positively correlated with precipitation, temperature, and GDP, and negatively correlated with the population density. Overall, climate-dominated aNPP increase and its contribution obviously increase in the past 40 year; while human activities were the main driving factor for the decrease of aNPP in China. This study explored the relative contributions of climate and human activities to aNPP changes in China over the past 40 years. It will provide scientific guidance for carbon sequestration increase of terrestrial vegetation and sustainable ecosystem development. (4) The regions that always maintained grassland in Tibetan Plateau, the grassland aNPP had a net increase of 58.5 Tg C during 1982–2000, and the contribution rates of climate and human activities were 35.7% and 64.3%, respectively. From 2001 to 2020, the net increase in grassland aNPP was 14.7 Tg C, and the contribution rates of climate and human activities were 223.4% and − 123.4%, respectively. (4) Before 2000, climate, human activities (except LUCC), and LUCC jointly promoted the increase in grassland aNPP. After 2000, climate change dominated the increase in grassland aNPP. Conversely, human activities (except LUCC) dominated the decrease in grassland aNPP. Therefore, for regions with human activity-dominated NPP decreases, effective strategies and policies should be precisely carried out, such as controlling livestock numbers, maintaining a grassland-livestock balance, providing differentiated ecological compensation incentivizing herdsmen's ecological protection initiative, implementing rotation grazing, house feeding and breeding, planting artificial grass, and strengthening publicity and education on ecological environment protection. These actions are crucial to strengthen the ecological security barrier and ensure sustainable regional development of the Tibetan Plateau.

References

Adams, B., White, A., & Lenton, T. M. (2004). An analysis of some diverse approaches to modelling terrestrial net primary productivity. *Ecological Modelling, 177*, 353–391.

Chang, X., Wang, S., Luo, C., Zhang, Z., Duan, J., Zhu, X., Lin, Q., & Xu, B. (2012). Responses of soil microbial respiration to thermal stress in alpine steppe on the Tibetan Plateau. *European Journal of Soil Science, 63*, 325–331.

Chen, H., Ju, P. J., Zhang, J., Wang, Y. Y., Zhu, Q. A., Yan, L., Kang, X. M., He, Y. X., Zeng, Y., Hao, Y. B., & Wang, Y. F. (2020). Attribution analyses of changes in alpine grasslands on the Qinghai-Tibetan Plateau. *Chinese Science Bulletin-Chinese, 65*, 2406–2418.

Chen, H., Zhu, Q., Peng, C., Wu, N., Wang, Y., Fang, X., Gao, Y., Zhu, D., & Yang, G. (2013). The impacts of climate change and human activities on biogeochemical cycles on the Qinghai-Tibetan Plateau. *Global Change Biology, 19*, 2940–2955.

Ciais, P., Reichstein, M., Viovy, N., Granier, A., Ogée, J., Allard, V., Aubinet, M., Buchmann, N., Bernhofer, C., Carrara, A., Chevallier, F., De Noblet, N., Friend, A. D., Friedlingstein, P., Grünwald, T., Heinesch, B., Keronen, P., Knohl, A., Krinner, G., et al. (2005). Europe-wide reduction in primary productivity caused by the heat and drought in 2003. *Nature, 437*, 529–533.

Dai, L. C., Fu, R. Y., Guo, X. W., Du, Y. G., Zhang, F. W., & Cao, G. M. (2022). Soil moisture variations in response to precipitation across different vegetation types on the northeastern Qinghai-Tibet Plateau. *Frontiers in Plant Science, 13*, 854152.

Du, M., Kawashima, S., Yonemura, S., Zhang, X., & Chen, S. (2004). Mutual influence between human activities and climate change in the Tibetan Plateau during recent years—ScienceDirect. *Global and Planetary Change, 41*, 241–249.

Fan, J., Wang, H., Chen, D., Zhang, W., & Wang, C. (2010). Discussion on sustainable urbanization in Tibet. *Chinese Geographical Science, 20*, 258–268.

Feng, Y. H., Zhu, J. X., Zhao, X., Tang, Z. Y., Zhu, J. L., & Fang, J. Y. (2019). Changes in the trends of vegetation net primary productivity in China between 1982 and 2015. *Environmental Research Letters, 14*, 124009.

Ge, N. N., Wei, X. R., Wang, X., Liu, X. T., Shao, M. A., Jia, X. X., Li, X. Z., & Zhang, Q. Y. (2019). Soil texture determines the distribution of aggregate-associated carbon, nitrogen and phosphorous under two contrasting land use types in the Loess Plateau. *Catena, 172*, 148–157.

Ge, W. Y., Deng, L. Q., Wang, F., & Han, J. Q. (2021). Quantifying the contributions of human activities and climate change to vegetation net primary productivity dynamics in China from 2001 to 2016. *Science of the Total Environment, 773*, 145658.

Gong, X. W., Li, Y. Q., Wang, X. Y., Zhang, Z. S., Lian, J., Ma, L., Chen, Y., Li, M. Y., Si, H. T., & Cao, W. J. (2022). Quantitative assessment of the contributions of climate change and human activities on vegetation degradation and restoration in typical ecologically fragile areas of China. *Ecological Indicators, 144*, 109536.

Huang, K., Zhang, Y. J., Zhu, J. T., Liu, Y. J., Zu, J. X., & Zhang, J. (2016). The influences of climate change and human activities on vegetation dynamics in the Qinghai-Tibet Plateau. *Remote Sensing, 8*, 876.

Huang, M. T., Piao, S. L., Ciais, P., Penuelas, J., Wang, X. H., Keenan, T. F., Peng, S. S., Berry, J. A., Wang, K., Mao, J. F., Alkama, R., Cescatti, A., Cuntz, M., De Deurwaerder, H., Gao, M. D., He, Y., Liu, Y. W., Luo, Y. Q., Myneni, R. B., et al. (2019). Air temperature optima of vegetation productivity across global biomes. *Nature Ecology & Evolution, 3*, 772–779.

Imhoff, M. L., Bounoua, L., DeFries, R., Lawrence, W. T., Stutzer, D., Tucker, C. J., & Ricketts, T. (2004). The consequences of urban land transformation on net primary productivity in the United States. *Remote Sensing of Environment, 89*, 434–443.

Jiang, L., Deng, X. Z., & Seto, K. C. (2013). The impact of urban expansion on agricultural land use intensity in China. *Land Use Policy, 35*, 33–39.

Li, Q., Zhang, C. L., Shen, Y. P., Jia, W. R., & Li, J. (2016). Quantitative assessment of the relative roles of climate change and human activities in desertification processes on the Qinghai-Tibet Plateau based on net primary productivity. *Catena, 147*, 789–796.

Liu, J., Li, S., Ouyang, Z., Tam, C., & Chen, X. (2008). Ecosystem services special feature: Ecological and socioeconomic effects of China's policies for ecosystem services. *Proceedings of the National Academy of Sciences, 105*, 9477–9482.

Lyu, X., Li, X. B., Dang, D. L., Dou, H. S., Wang, K., Gong, J. R., Wang, H., & Liu, S. L. (2022). A Perspective on the impact of grassland degradation on ecosystem services for the purpose of sustainable management. *Remote Sensing, 14*, 5120.

Macias-Fauria, M. (2018). Satellite images show China going green. *Nature, 553*, 411–413.

Mayor, J. R., Sanders, N. J., Classen, A. T., Bardgett, R. D., Clément, J. C., Fajardo, A., Lavorel, S., Sundqvist, M. K., Bahn, M., Chisholm, C., Cieraad, E., Gedalof, Z., Grigulis, K., Kudo, G., Oberski, D. L., & Wardle, D. A. (2017). Elevation alters ecosystem properties across temperate treelines globally. *Nature, 542*, 91.

Naeem, S., Zhang, Y. Q., Tian, J., Qamer, F. M., Latif, A., & Paul, P. K. (2020). Quantifying the impacts of anthropogenic activities and climate variations on vegetation productivity changes in China from 1985 to 2015. *Remote Sensing, 12*, 1113.

Nemani, R. R., Keeling, C. D., Hashimoto, H., Jolly, W. M., Piper, S. C., Tucker, C. J., Myneni, R. B., & Running, S. W. (2003). Climate-driven increases in global terrestrial net primary production from 1982 to 1999. *Science, 300*, 1560–1563.

Niu, K. C., Choler, P., Zhao, B., & Du, G. Z. (2009). The allometry of reproductive biomass in response to land use in Tibetan alpine grasslands. *Functional Ecology, 23*, 274–293.

Ouyang, Z., Zheng, H., Yi, X., Polasky, S., Liu, J., Xu, W., & Wang, Q. (2016). Improvements in ecosystem services from investments in natural capital. *Science, 352*, 1455–1459.

Peng, S., Piao, S., Wang, T., Sun, J., & Shen, Z. (2009). Temperature sensitivity of soil respiration in different ecosystems in China. *Soil Biology & Biochemistry, 41*, 1008–1014.

Piao, S. L., Fang, J. Y., Zhou, L. M., Zhu, B., Tan, K., & Tao, S. (2005). Changes in vegetation net primary productivity from 1982 to 1999 in China. *Global Biogeochemical Cycles, 19*, 1029.

Piao, S. L., Wang, X. H., Park, T., Chen, C., Lian, X., He, Y., Bjerke, J. W., Chen, A., Ciais, P., Tommervik, H., Nemani, R. R., & Myneni, R. B. (2020). Characteristics, drivers and feedbacks of global greening. *Nature Reviews Earth & Environment, 1*, 14–27.

Rafique, R., Zhao, F., de Jong, R., Zeng, N., & Asrar, G. R. (2016). Global and regional variability and change in terrestrial ecosystems net primary production and NDVI: a model-data comparison. *Remote Sensing, 8*, 8030177.

Rawls, W. J., Pachepsky, Y. A., Ritchie, J. C., Sobecki, T. M., & Bloodworth, H. (2003). Effect of soil organic carbon on soil water retention. *Geoderma, 116*, 61–76.

Sha, Z. Y., Bai, Y. F., Li, R. R., Lan, H., Zhang, X. L., Li, J. T., Liu, X. F., Chang, S. J., & Xie, Y. C. (2022). The global carbon sink potential of terrestrial vegetation can be increased substantially by optimal land management. *Communications Earth & Environment, 3*, 8.

Song, X., Yang, G., Yan, C., Duan, H., Liu, G., & Zhu, Y. (2009). Driving forces behind land use and cover change in the Qinghai-Tibetan Plateau: A case study of the source region of the Yellow River, Qinghai Province, China. *Environmental Earth Sciences, 59*, 793–801.

Song, Y., & Ma, M. G. (2011). A statistical analysis of the relationship between climatic factors and the normalized difference vegetation index in China. *International Journal of Remote Sensing, 32*, 3947–3965.

Tong, X. W., Brandt, M., Yue, Y. M., Horion, S., Wang, K. L., De Keersmaecker, W., Tian, F., Schurgers, G., Xiao, X. M., Luo, Y., Chen, C., Myneni, R., Shi, Z., Chen, H. S., & Fensholt, R. (2018). Increased vegetation growth and carbon stock in China karst via ecological engineering. *Nature Sustainability, 1*, 44–50.

Wang, H., Liu, G. H., Li, Z. S., Zhang, L. W., & Wang, Z. Z. (2020). Processes and driving forces for changing vegetation ecosystem services: Insights from the Shaanxi Province of China. *Ecological Indicators, 112*, 106105.

Wang, J., Guo, N., Cai, D. H., & Deng, Z. Y. (2009). The effect evaluation of the program of restoring grazing to grasslands in Maqu County. *Acta Ecologica Sinica, 29*, 1276–1284.

Wang, L. H., Tian, F., Wang, Y. H., Wu, Z. D., Schurgers, G., & Fensholt, R. (2018). Acceleration of global vegetation greenup from combined effects of climate change and human land management. *Global Change Biology, 24*, 5484–5499.

Wu, L. H., Wang, S. J., Bai, X. Y., Tian, Y. C., Luo, G. J., Wang, J. F., Li, Q., Chen, F., Deng, Y. H., Yang, Y. J., & Hu, Z. Y. (2020). Climate change weakens the positive effect of human activities on karst vegetation productivity restoration in southern China. *Ecological Indicators, 115*, 106392.

Wu, N., & Liu, Z. (1998). Probing into causes of geographical pattern of subalpine vegetation on the eastern Qinghai-Tibetan Plateau. *Chinese Journal of Apllied Environment, 4*, 290–297.

Wu, Y. Y., & Wu, Z. F. (2023). NPP variability associated with natural and anthropogenic factors in the tropic of cancer transect, China. *Remote Sensing, 15*, 1091.
Xu, C., Zhang, L. B., Du, J. Q., Guo, Y., Wu, Z. F., Xu, Y. D., Li, F., & Wang, F. Y. (2013). Impact of alpine meadow degradation on soil water conservation in the source region of three rivers (in Chinese). *Acta Ecolologica Sinica, 33*, 2388–2399.
Xu, D. Y., Kang, X. W., Zhuang, D. F., & Pan, J. J. (2010). Multi-scale quantitative assessment of the relative roles of climate change and human activities in desertification—A case study of the Ordos Plateau, China. *Journal of Arid Environments, 74*, 498–507.
Xu, H. J., Wang, X. P., & Zhang, X. X. (2016). Alpine grasslands response to climatic factors and anthropogenic activities on the Tibetan Plateau from 2000 to 2012. *Ecological Engineering, 92*, 251–259.
Xu, Z. Z., & Zhou, G. S. (2006). Combined effects of water stress and high temperature on photosynthesis, nitrogen metabolism and lipid peroxidation of a perennial grass Leymus chinensis. *Planta, 224*, 1080–1090.
Yan, Z. L., & Wu, N. (2005). Rangeland privatization and its impacts on the Zoige wetlands on the Eastern Tibetan Plateau. *Journal of Mountain Science, 2*, 105–115.
Yu, D. Y., Shi, P. J., Han, G. Y., Zhu, W. Q., Du, S. Q., & Xun, B. (2011). Forest ecosystem restoration due to a national conservation plan in China. *Ecological Engineering, 37*, 1387–1397.
Zhou, W., Gang, C., Zhou, F., Li, J., Dong, X., & Zhao, C. (2015). Quantitative assessment of the individual contribution of climate and human factors to desertification in northwest China using net primary productivity as an indicator. *Ecological Indicators, 48*, 560–569.
Zhou, W., Wang, T., Xiao, J. Y., Wang, K. M., Yu, W. P., Du, Z. P., Huang, L., & Yue, T. X. (2024). Grassland productivity increase was dominated by climate in Qinghai-Tibet Plateau from 1982 to 2020. *Journal of Cleaner Production, 434*, 140144.
Zhou, W., Yang, H., Huang, L., Chen, C., Lin, X. S., Hu, Z. J., & Li, J. L. (2017). Grassland degradation remote sensing monitoring and driving factors quantitative assessment in China from 1982 to 2010. *Ecological Indicators, 83*, 303–313.
Zhu, W., Pan, Y., He, H., Yu, D., & Hu, H. (2006). Simulation of maximum light use efficiency for some typical vegetation types in China. *Chinese Science Bulletin, 51*, 457–463.

Chapter 7
Accounting and Dynamics Assessment of Ecological Functions

Abstract The three rivers source region, known as the "Water Tower of China," plays a crucial role in ensuring regional ecological security through its dynamic ecological barrier functions. This study quantitatively evaluated the evolution of four core functions—carbon sequestration, water conservation, soil and water conservation, and biodiversity—using land use data from 2000 and 2018, combined with multi-source geographic information and meteorological data, through the InVEST model and principal component analysis. The results showed that: (1) Total carbon storage increased significantly from 753.7 Tg in 2000 to 781.2 Tg in 2018, with an increment of 27.5 Tg. Forest land (49.689%), grassland (27.709%), and unused land (18.653%) constitute the core carbon reservoirs, maintaining a spatial pattern of "higher in the southeast and lower in the northwest." (2) The capacity for water conservation showed remarkable improvement, with average runoff increasing from 137 to 229 mm (a 67.2% growth). The spatial distribution pattern evolved from fragmented to concentric circles radiating outward from central areas (Nangqian County and Yushu City). (3) Soil and water conservation exhibited significant regional heterogeneity: Although erosion was reduced in local areas like the Tongtian River Basin, the total bare soil erosion increased from 1,434,479.405 million tons to 1,650,740.271 million tons, while total soil erosion rose from 215,771.375 million tons to 295,158.724 million tons, indicating intensified erosion pressure in human-affected southeastern areas. (4) There was a steady improvement in biodiversity maintenance: The average habitat quality index rose from 0.891 to 0.907, while the average relative habitat degradation level decreased from 0.008 to 0.007. However, degradation hotspots around roads and urban areas persist. The comprehensive ecological barrier function index indicates that ecosystem services in the three rivers source region showed positive trends between 2000 and 2018, demonstrating the effectiveness of ecological engineering initiatives and providing data support for precision ecological management.

Keywords Ecological function · Carbon storage · Soil and water conservation · Habitat quality

© The Author(s), under exclusive license to Springer Nature Singapore Pte Ltd. 2026

W. Zhou et al., *Carbon Sink Estimation and Evaluation of Ecological Function Changes on the Tibetan Plateau*, https://doi.org/10.1007/978-981-95-8664-6_7

7.1 Introduction

Nestled in the heart of the Qinghai-Tibet Plateau, the three rivers source region is renowned as the "Water Tower of China" due to its abundant rivers, lakes, and marshes, along with extensive glaciers. This ecological paradise is the origin of the Yangtze, Yellow, and Lancang Rivers and features the world's largest natural high-altitude wetland, which highlights its crucial ecological importance.

However, since the 1980s, economic development has resulted in the destruction of over 16 million mu (approximately 1.07 million hectares) of grasslands. Soil erosion has led to severe desertification and degradation of pastures, reduced water retention capacity, and endangered biodiversity. These environmental alterations have set off a chain reaction that affects ecosystems, economies, and security. The decline of Maduo County, a core area of the region, serves as a stark example of this ecological crisis.

Ecological civilization construction is related to people's well-being, and the protection of the ecological environment serves as a crucial foundation for China to achieve sustainable development (Liu, 2012).

As the 3S technologies—remote sensing (RS), geographic information systems (GIS), and global positioning systems (GPS)—continue to advance, they have become indispensable in information processing and dynamic monitoring across numerous fields, particularly in the detection and evolution analysis of regional ecological environments. Nevertheless, geographic information data, especially the latest remote sensing technologies, have been underutilized in monitoring ecological restoration measures and the spatiotemporal evolution of ecological functions in the Three Gorges Region.

This study utilizes multi-source remote sensing data fusion technology to conduct a comprehensive analysis of remote sensing data in the Three Gorges Region. It also employs geographic information models and ecological service value assessment models to provide a scientific basis for spatiotemporal monitoring during ecological restoration processes.

Several models are available for assessing the value of ecosystem services, such as the Integrated Valuation of Ecosystem Services and Trade-offs (InVEST) model, the Ecological Footprint Model (EFM), the CITYgreen model, and the Ecological Value at Risk (EVR) model.

Scholars have extensively utilized remote sensing technology to evaluate ecosystem service values. In land use studies, Petit and Lambin (2001) pioneered integrated monitoring through map synthesis and spatial analysis of land cover. For assessing ecosystem moisture status and vegetation productivity, Ivits et al. (2014) employed the Soil–Plant Interactions Index (SPEI) as a drought indicator and analyzed vegetation growth cycles via NDVI time series. Wang et al. (2019) conducted dynamic monitoring of the Central Plains Urban Agglomeration's ecological value from 2001 to 2013, achieving pixel-level analysis of ecosystem service value (ESV) through spatial scale effects in multi-source remote sensing models.

In modeling the carbon cycle within terrestrial ecosystems, Dixon's research uncovered complex future trends in forest carbon cycles, which are associated with global climate and land use changes. Some predictions indicate that forest ecosystems could function as carbon sinks (Dixon et al., 1994). Other studies have utilized LiDAR to accurately quantify biomass and carbon dynamics in coniferous forests. Further, they demonstrated that even mature stands within these forests have the potential to act as carbon sinks. With respect (Hudak et al., 2012) to modeling vegetation ecosystem productivity or the carbon cycle, Zhang (2006) wetland ecosystem carbon cycle model has been developed, offering a robust framework for terrestrial carbon simulations.

For ecological restoration in degraded watersheds, estimate the economic costs associated with the increasing demands for ecosystem services (Loomis et al., 2000). For soil condition evaluation, Ali and his colleagues integrated data from soil and water conservation interventions in Ethiopia's Tana—Belso River Basin (2009–2016) with soil moisture data extracted from satellite imagery to assess the impact of these conservation measures (Ali et al., 2020).

The biodiversity index serves as a crucial indicator of ecological changes. Obtaining various diversity metrics often requires complex calculations. Currently, Microsoft Excel functions are widely used to calculate biodiversity indices. Additionally, some researchers use geographical methods to study biodiversity indices. For example, Herrmann et al. (2003) used GIS modeling tools to simultaneously reflect multiple factors through scenario analysis, revealing the close correlation between biodiversity changes across different regions.

The InVEST model primarily simulates ecological indicators to address current shortages in soil and water resources. Modules such as carbon storage, water yield, sediment transport ratio, and habitat quality are widely applied. In the assessment of ecological services and values, Redhead et al. validated parameter variations and error sensitivity within the aquatic ecosystem service module using the InVEST model (Redhead et al., 2016). Goldstein et al. used the InVEST model to evaluate the ecological values of Hawaii's central islands, offering insights for optimizing agricultural irrigation systems and production models (Goldstein et al., 2012). Researchers have also assessed the ecological service values in the three rivers source region. Some studies have simulated spatiotemporal changes in water source conservation, grassland ecological compensation, and water supply services in the three rivers source region using the InVEST model (Cheng, 2019; Pan et al., 2013).

Current ecological research on human–land relationships in the three rivers source region primarily focuses on three key areas: environmental ecology and natural elements, sustainable development, and ecological conservation. Studies that utilize multi-source data on human activities and vegetation coverage have emerged as a research hotspot (Wei et al., 2020). However, there is still a lack of comprehensive evaluation and dynamic monitoring of ecological barrier functions based on multi-source data. Additionally, there is insufficient analysis of the evolution of functional indicators and the development of integrated evaluation systems for these critical ecological barriers.

Based on this, the paper evaluates and monitors the ecological barrier functions of the three rivers source region. By integrating remote sensing data, field survey data, and literature data, and utilizing multiple data processing and analysis platforms, the study employs the four evaluation modules of the InVEST terrestrial ecosystem assessment model to simulate and obtain assessment results for carbon sequestration and fixation, water conservation, soil and water conservation, and biodiversity in the three rivers source region. A functional index system for the ecological barrier is constructed to monitor the effectiveness of ecological restoration in the region and provide a reference for the quantitative evaluation of the driving forces behind ecological barrier functions.

7.2 Data Sources and Methods

7.2.1 Overview of the Three Rivers Source Region

The three rivers source region is located in the western part of China, with an average altitude ranging from 3500 to 4800 m. It is the core area of the world's roof—the Qinghai—Tibet Plateau. Situated in the southern part of Qinghai Province, it is the birthplace of the Yangtze River, the Yellow River, and the Lancang River (also known as the Mekong River). These rivers have nurtured Chinese civilization and the long-standing history of the Indochina Peninsula. Geographically, it spans from north latitude 31°39′ to 36°12′ and east longitude 89°45′ to 102°23′, covering 16 counties. With a total area of 302,500 km^2, it accounts for approximately 43% of Qinghai Province's land area. The current population is 556,000, over 90% of whom are Tibetan. The three rivers region is mainly mountainous, with continuous mountain ranges, high elevations, and complex terrain. The altitude varies from 3335 to 6564 m. The lowest point is along the Jinsha River in the southeastern part of Yushu Tibetan Autonomous Prefecture, with an average altitude of about 4400 m. The region's topography is mainly characterized by high mountains at altitudes of 4000 to 5800 m. The main mountain ranges include the Eastern Kunlun Mountains and their branches, the Amnye Machen Mountains, the Bayan Har Mountains, and the Tanggula Mountains. The climate of the three rivers region belongs to the Qinghai–Tibet Plateau climate system, featuring a typical plateau continental climate. It has alternating cold and hot seasons, distinct dry and wet seasons, small annual temperature variations, large daily temperature differences, long hours of sunshine, intense radiation, and a climate without the distinct features of the four seasons.

7.2.2 Main Data Sources and Processing

This study employed multi-source data, including topography, meteorology, soil characteristics, human activities, and additional information. All datasets were standardized to the WGS_1984_Albers coordinate system and resampled at a 1 km resolution to ensure comparability. Table 7.1 outlines the sources and characteristics of the primary data.

Table 7.1 Description of main data sources

Data	Data content	Resolution (km)	Source
Landform	Geomorphic type	1	Center for Resource and Environmental Sciences and Data (http://www.resdc.cn)
	Digital elevation model	1	Center for Resource and Environmental Sciences and Data (http://www.resdc.cn)
	Slope, slope direction	1	Obtained from DEM
Meteorological phenomena	Monthly precipitation and potential evapotranspiration	1	National Earth System Science Data Platform (http://loess.geodata.cn)
Soil	Soil texture (sand, silt, clay)	1	Center for Resource and Environmental Sciences and Data (http://www.resdc.cn)
	Agrotype	1	Center for Resource and Environmental Sciences and Data (http://www.resdc.cn)
	Permafrost type		National Tibetan Plateau Science Data Center
Human activities	Road network	1	OpenStreetMap http://download.geofabrik.de/asia.html
	Density of population	1	Center for Resource and Environmental Sciences and Data (http://www.resdc.cn)
	Land use	1	Institute of Geographic Sciences and Natural Resources Research, Chinese Academy of Sciences
Basic geography	Administrative area	1	National Center for Basic Geographic Information (http://www.ngcc.cn)
Model parameter	Soil erodability factor	1	National Center for Earth System Science Data (http://loess.geodata.cn)
	Maximum root depth in soil		Canadell et al. literature

Table 7.2 Maximum root burial depth of soil in different land use types

Land use type	Land type code	Maximum root system burial depth (mm)
Plough	1	2100
Forest land	2	7000
Meadow	3	2600
Waters	4	10
Construction	5	10
Unutilized land	6	10

7.2.2.1 Data Processing Flow

(1) Topographic Data: Slope and aspect were extracted from DEM data using ArcGIS 10.4. The aspect was reclassified into nine categories: north, northeast, east, southeast, south, southwest, west, northwest, and flat. Landform types were categorized into six classes: plains, terraces, hills, low-mountainous areas, medium-mountainous areas, and high-mountainous areas.
(2) Meteorological Data: Monthly precipitation and potential evapotranspiration data were obtained and processed in ArcMap to generate annual averages, serving as indicators for annual precipitation and annual potential evapotranspiration.
(3) Soil Data: Soil Texture: The percentage content of sand, silt, and clay is divided into five grades (0–20%, 20–40%, 40–60%, 60–80%, 80–100%). Soil Type: According to the "soil genesis classification" system, the soil in the area is divided into nine soil classes. Type of Frozen Soil: Frozen soil data from different periods were integrated, and the frozen soil types were classified into two types: perennial frozen soil and seasonal frozen soil.
(4) Human Activity Data: Road Network: Using OpenStreetMap vector data, road network density and habitat threat factor raster data were calculated. Land Use: The original 23 secondary land types were reclassified into six primary types: cultivated land, woodland, grassland, water area, construction land, and unused land.
(5) Model Parameters: The maximum root burial depth in soil is based on the research findings of Canadell et al. (Table 7.2), while the soil erodibility factor (K) directly adopts the 1 km resolution physical dataset.

7.2.2.2 Data Integration and Sampling

After processing all factor data to a 1 km resolution, systematic sampling was conducted using 3 km × 3 km fishery station vector data. Following the removal of samples with empty values, 39,432 valid geographic samples were obtained for subsequent modeling and analysis.

7.2.3 *Research Methods*

7.2.3.1 InVEST Model

This study utilized the InVEST model's modules for carbon storage, water yield, sediment transport ratio, and biodiversity to simulate the ecological barrier functions and benefits in the three rivers source region.

(1) **Carbon Storage and Fixation Module**

The principle involves calculating the carbon stock and carbon sequestration of each raster by using the land use raster and all input carbon pool data (above ground, below ground, soil, dead organic matter). The calculation formula is as follows:

$$C_{\text{tot}} = C_{\text{above}} + C_{\text{below}} + C_{\text{soil}} + C_{\text{dead}} + C_{\text{tree}} \tag{7.1}$$

where C_{tot} is the total carbon storage, C_{above} is the aboveground carbon storage, C_{below} is the underground carbon storage, C_{soil} is the soil carbon storage, and C_{dead} is the dead organic matter carbon storage (Table 7.3).

(2) **Water Production Module**

By inputting processed data—including maximum root depth, annual precipitation, plant available water, annual potential evapotranspiration, land use raster data, watershed vector data, biophysical coefficient table, and seasonal constant Z—into the model, we can obtain the actual evapotranspiration ratio, actual evapotranspiration, and water yield raster data within the study area. The formula is as follows:

$$Y(x) = \left(\left[1 + \left(\frac{\text{PET}(X)}{P(x)}\right)^{\omega}\right]^{\frac{1}{\omega}} - \frac{\text{PET}(X)}{P(x)}\right) \times P(x) \tag{7.2}$$

In the formula, $Y(x)$ represents the annual water yield of the landscape type grid x, PET(x) denotes the potential evapotranspiration, $P(x)$ signifies the annual precipitation of the grid unit, and ω is a non-physical parameter related to natural climate

Table 7.3 Four carbon density values (t/hm^2) for different land use types

Land use type	Land type code	Land biomass	Subsurface biomass	Soil carbon density	Dead organic matter
Plough	1	3.15	8.07	10.84	1.1
Forest land	2	4.24	11.59	12.33	5.8
Meadow	3	3.46	8.65	10	5
Waters	4	0	0	0	1.2
Construction	5	0	0	7.8	0
Unutilized land	6	1.2	0	7.1	0

and soil properties, which describes the ratio of the plant's annual water demand to the precipitation.

(3) **Sediment Transport Ratio**

The soil and water conservation output data consists of four parameters: bare soil erosion volume, total soil erosion, potential total soil erosion, and sediment deposition index. These parameters are calculated using a Digital Elevation Model (DEM), a precipitation erosivity factor (R), a soil erodability factor (K), land use data, watershed data, biophysical coefficient tables, and parameters including catchment area threshold, kb, IC0, and SDRmax.

(4) **Biodiversity Module**

The relative habitat degradation level of the current landscape type is assessed using a scale where 1 indicates high degradation and 0 indicates low degradation, with non-habitat areas scored as 0. Concurrently, the habitat quality of the current landscape type is evaluated, with higher values denoting better quality, and non-habitat areas also scored as 0. The quality score is dimensionless and does not represent specific biodiversity measurements. These evaluations are conducted using the following formula:

$$Q_x = \mathrm{H}_j\left(1 - \left(\frac{D_x^2}{D_x^2 + k^2}\right)\right) \tag{7.3}$$

In the formula: Q_x represents the habitat quality of land use classification raster x, D_x indicates the habitat stress level of the land use classification raster, k is the semi-saturation constant, and H_j denotes the habitat adaptability of the land use classification data. The biodiversity module of the InVEST model requires the following inputs: land use classification data, land use classification sensitivity data for each threat factor, spatial data on the distribution density of each threat, and spatial location data of protected areas.

7.2.3.2 Principal Component Analysis

Principal component analysis (PCA) has increasingly been applied in ecological assessment studies that involve multi-indicator screening. Given the multitude of ecological indicators, PCA emerges as a practical approach for evaluating and consolidating multi-indicator performance in comprehensive ecological benefit assessments. Therefore, this study employs PCA to construct an indicator system.

Scientifically selecting indicators is crucial for evaluating the credibility of assessment results. After reviewing extensive literature, we identified 10 key indicators: NDVI, NPP, total carbon storage, relative habitat degradation, habitat quality, total soil erosion, water yield, annual average precipitation, and annual average temperature. The principal components derived from the analysis were interpreted as criterion indicators, forming an indicator system with the ecological barrier function-efficiency index as the target layer, as shown in Table 7.4.

The total variance explained and component matrix of the index data were obtained using SPSS software, as shown in Tables 7.5 and 7.6.

To ensure the extraction of four principal components annually over a span of two years, the eigenvalue extraction threshold was set at 0.8. The cumulative contribution rate of the first four components reached 78.206%, surpassing the 70% threshold and effectively explaining the selected 10 indicators. In 2018, the cumulative contribution rate of these four components increased to 79.411%. Statistical tests indicated KMO values of 0.739 and 0.753, respectively, signifying the data's suitability for principal component analysis. These components essentially represent the original

Table 7.4 Comprehensive evaluation index system for ecological barrier functions in the three rivers source region

Target layer (I)	Policy level (A)	Indicator layer (B)
Ecological barrier function and benefit index	A1	NDVI (B1)
		NPP (B2)
	A2	Total carbon stock (B3)
		Relative Habitat Degradation (B4)
		Habitat quality (B5)
	A3	Soil erosion intensity (B6)
		Water yield (B7)
	A4	Slope (B8)
		Annual Precipitation (B9)
		Annual average temperature (B10)

Table 7.5 Total variance explained by principal component analysis in 2000

Ingredient	Initial eigenvalues			Extract sum of squares of loads		
	Amount to (λ)	Variance percentage	Accumulate %	Amount to	Variance percentage	Accumulate %
1	3.668	36.678	36.678	3.668	36.678	36.678
2	1.957	19.569	56.248	1.957	19.569	56.248
3	1.322	13.215	69.463	1.322	13.215	69.463
4	0.874	8.743	78.206	0.874	8.743	78.206
5	0.678	6.784	84.991			
6	0.571	5.708	90.699			
7	0.359	3.594	94.293			
8	0.276	2.761	97.054			
9	0.222	2.222	99.276			
10	0.072	0.724	100			

Table 7.6 Total variance explained by principal component analysis in 2018

Ingredient	Initial eigenvalues			Extract sum of squares of loads		
	Amount to (λ)	Variance percentage	Accumulate %	Amount to	Variance percentage	Accumulate %
1	3.934	39.338	39.338	3.934	39.338	39.338
2	1.730	17.297	56.634	1.730	17.297	56.634
3	1.161	11.611	68.246	1.161	11.611	68.246
4	1.117	11.165	79.411	1.117	11.165	79.411
5	0.684	6.840	86.251			
6	0.605	6.050	92.301			
7	0.367	3.668	95.969			
8	0.200	2.004	97.973			
9	0.125	1.253	99.227			
10	0.077	0.773	100			

10 indicators in the comprehensive evaluation of ecological barrier functions in the three rivers source region.

Based on the correlation coefficients of each evaluation index in the principal component matrix and the loadings of each component in the component matrix, the results of the two-year principal component analysis are comprehensively analyzed to identify the evaluation indexes with high correlation with the principal components, and the qualitative analysis of each principal component is completed.

As indicated in Tables 7.7 and 7.8, indicators closely associated with carbon sequestration and oxygen release, such as Net Primary Productivity (NPP), Normalized Difference Vegetation Index (NDVI), annual average temperature, carbon storage, and annual precipitation, exhibit high loadings on the first principal component. This suggests a significant positive correlation, thus allowing A1 to be interpreted as representing the carbon sequestration and oxygen release function.

The second principal component displays high loadings for habitat quality and relative habitat degradation, both of which are correlated with species diversity. Consequently, A2 can be interpreted as a species conservation function.

The third principal component demonstrates a strong positive correlation with water yield, suggesting that A3 can be interpreted as representing water source conservation.

The fourth principal component exhibits a strong correlation with slope and total soil erosion, while showing a negative correlation with annual precipitation and water yield. Therefore, this principal component, A4, represents soil and water conservation functions.

Finally, it was decided to use these four criteria to replace the original 10 indicators to comprehensively evaluate the ecological barrier function of the source area of the three rivers.

Table 7.7 2000 Principal component analysis component matrix

Metric	Ingredient			
	V1	V2	V3	V4
Zscore (B3)	0.737	– 0.547	0.219	– 0.018
Zscore (B5)	0.684	– 0.593	0.255	– 0.060
Zscore (B4)	0.039	– 0.553	0.617	– 0.011
Zscore (B7)	– 0.036	0.606	0.528	– 0.495
Zscore (B6)	– 0.008	0.540	0.562	0.321
Zscore (B1)	0.859	0.147	– 0.187	0.054
Zscore (B8)	0.460	0.320	0.247	0.615
Zscore (B10)	0.709	0.204	– 0.280	0.102
Zscore (B9)	0.740	0.362	0.112	– 0.330

Note The table includes NDVI (B1), NPP (B2), total carbon storage (B3), relative habitat degradation (B4), habitat quality (B5), soil erosion intensity (B6), water yield (B7), slope (B8), annual precipitation (B9), and annual temperature (B10).

Table 7.8 2018 PCA component matrix

Metric	Ingredient			
	V1	V2	V3	V4
Zscore (B5)	0.637	0.687	0.023	0.033
Zscore (B4)	0.013	0.634	– 0.208	0.487
Zscore (B7)	0.093	0.091	0.908	– 0.198
Zscore (B3)	0.699	0.623	0.035	0.044
Zscore (B2)	0.890	– 0.127	0.078	– 0.150
Zscore (B1)	0.870	– 0.139	0.085	– 0.173
Zscore (B9)	0.806	– 0.359	– 0.239	0.040
Zscore (B10)	0.761	– 0.388	– 0.246	0.093

Note The table includes NDVI (B1), NPP (B2), total carbon storage (B3), relative habitat degradation (B4), habitat quality (B5), soil erosion intensity (B6), water yield (B7), slope (B8), annual precipitation (B9), and annual temperature (B10)

$$U_i = \frac{V_i}{\sqrt{\lambda_i}} \tag{7.4}$$

$$A_1 = B'U_1 \quad A_2 = B'U_2 \quad A_3 = B'U_3 \quad A_4 = B'U_4 \tag{7.5}$$

where V1 denotes the component matrix, λ represents the eigenvalues of the principal components, and B1′ to B8′ are the Z-score standardized score vectors of the original variables B, with B′ = (B1′, B2′, B3′, B4′, B5′, B6′, B7′, B8′, B9′, B10′), U1, U2, U3, and U4 represent the eigenvectors of the first to fourth principal components, respectively. Using Excel's MMULT array formula, we calculated the values of A1,

Table 7.9 Partial values of the four principal components in 2000 and 2018

A particular year	Geographic sample	A1 (Carbon sequestration and oxygen release)	A2 (Species Conservation)	A3 (Water conservation)	A4 (Soil conservation)
2000	1	2.599	7.195	7.666	2.983
	2	1.033	1.755	2.813	– 0.290
	3	– 0.379	1.334	– 1.337	0.710
	4	2.387	1.157	0.911	1.150
	5	1.455	3.671	– 0.007	0.594
	6	– 0.447	1.185	– 1.530	0.284
	7	– 1.468	– 2.132	– 1.234	1.009
	8	3.572	– 0.025	– 1.620	– 0.043
	9	– 4.425	1.401	– 3.005	2.173
2018	1	2.781	– 0.082	0.066	0.073
	2	0.381	– 0.628	0.224	0.461
	3	1.224	– 0.024	0.006	0.340
	4	2.077	– 0.091	0.131	0.781
	5	3.329	– 0.165	0.172	0.647
	6	– 0.686	– 0.558	– 0.335	0.118
	7	– 4.900	– 0.546	– 0.073	– 0.628
	8	2.826	– 0.173	– 0.013	– 0.263
	9	– 2.869	– 0.991	– 0.413	0.073

A2, A3, and A4 for the variables. The ecological barrier functions of the three rivers source region were computed using the same method for 2000 and 2018, with the results presented in Table 7.9.

7.2.3.3 Geodetector Method

To quantitatively analyze the driving factors influencing the ecological barrier function of the Three Gorges Region, explain regional changes, and clarify the interrelationships among these factors, this study employs the geodetector method to investigate the driving forces behind the ecological benefits of the three rivers source region. The geodetector comprises four modules: factor detection, risk detection, interaction detection, and ecological detection. The factor detection and interaction detection modules are utilized in this study.

(1) Factor Detector

The q-value quantifies the spatial dissimilarity between attributes I and Ai, and measures how much a factor Xi explains their spatial variation. With a range of [0,1], a higher q-value indicates stronger explanatory power of Xi on I or Ai.

A simple transformation of the q-value satisfies the non-central F distribution.

$$F = \frac{N-L}{L-1}\frac{q}{1-q} \sim F(L-1, N-L; \lambda) \tag{7.6}$$

$$\lambda = \frac{1}{\sigma^2}\left[\sum_{h=1}^{L} \overline{I}_h^{2} - \frac{1}{N}\left(\sum_{h=1}^{L}\sqrt{N_h}\overline{I}_h\right)^2\right] \tag{7.7}$$

In the formula, λ is the non-central parameter and I_h is the mean of layer h. As per Eq. (7.7), the geodetector software includes a built-in function to test the significance of the q-value.

(2) **Interaction Detection**

To analyze the interaction effects between different factors Xs, we assess whether the interaction between factors X1 and X2 has a positive or negative impact on the dependent variable I or Ai, or if these factors act independently without common interaction effects. The evaluation method involves first calculating the q-values of factors X1 and X2 on I or Ai separately (q(X1) and q(X2)), then computing their interaction q-value (q(X1 ∩ X2)), and finally comparing q(X1), q(X2), and q(X1 ∩ X2) (Table 7.10).

(3) **Geodetector Application**

Based on the ecological benefit index derived from principal component analysis, this study utilizes various analytical modules of the Geographic Detector to examine the factors influencing the ecological benefits of the three rivers source region. Through quantitative analysis of these factors, the research determines their relative significance and uncovers the interaction mechanisms among the key factors.

Using ArcGIS 10.4, we created 3 km × 3 km raster and vector point datasets for the three river administrative region. The sampling tool was used to extract processed indicator data, and principal component analysis (PCA) was applied to the input

Table 7.10 Types of interaction between two independent variables and the dependent variable

Criterion	Interaction
q(X1 ∩ X2) < Min(q(X1),q(X2))	Nonlinear attenuation
Min(q(X1),q(X2)) < q(X1 ∩ X2) < Max(q(X1),q(X2))	Single-factor nonlinear attenuation
Max(q(X1),q(X2)) < q(X1 ∩ X2) < q(X1) + q(X2)	Dual factor enhancement
q(X1 ∩ X2) = q(X1) + q(X2)	Independence
q(X1) + q(X2) < q(X1 ∩ X2)	Nonlinear enhancement

Table 7.11 Driving force detection index system

Discovery field (Z)	Specific indicators (X)
Natural environment (Z1)	Altitude (X1)
	Slope (X2)
	Slope direction (X3)
	Terrain type (X4)
	Annual precipitation (X5)
	Annual average temperature (X6)
	Annual average relative humidity (X7)
	Sand content (X8)
	Silt content (X9)
	Clay content (X10)
	Soil type (X11)
	Permafrost type (X12)
Human activities (Z2)	Road network density (X13)
	Land use type (X14)
	Population density (X15)

dependent variable to obtain four PCA scores and the ecological barrier function efficiency coefficient (I).

Given that the geodetector generates overflow errors when the sample size exceeds 32,766, the study opted for a spatial resolution of 1 km for the research area. Utilizing Excel's filtering function, 12,304 geographic samples were systematically chosen from 39,432 geographic units for geodetector analysis. The data input format is illustrated in Table 7.11.

7.3 Results

7.3.1 Ecological Benefit Accounting

7.3.1.1 Carbon Storage Simulation

The carbon storage module of the INVEST model yields two outputs: a web-based summary of carbon storage data and the latest raster data. The three rivers source region registered a total carbon storage of 753.7 Tg in 2000, which rose to 781.2 Tg by 2018. Analysis of the two-year comparison indicates that although total carbon storage has increased, its spatial distribution has not significantly changed. The distribution pattern of carbon storage in 2018 shows a gradual decline from southeast

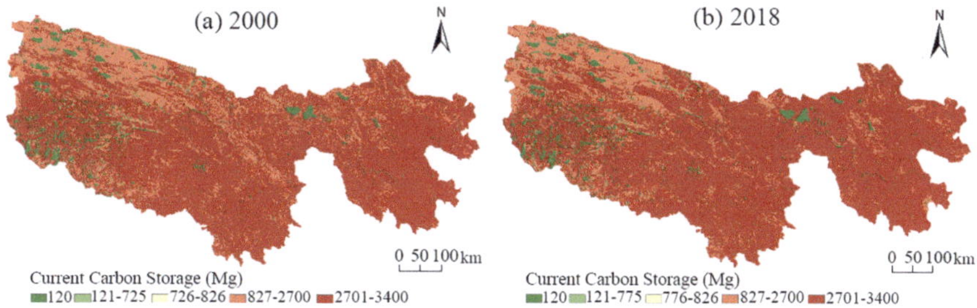

Fig. 7.1 Spatial distribution of carbon storage in **a** 2000 and **b** 2018

to northwest, with the lowest values centered in Maduo County and the Tanggula Mountains (Fig. 7.1).

7.3.1.2 Water Production Simulation

The data indicates that the water yield distribution in the three rivers region varied from 0 to 578.942 mm/km^2, with an average of 137 mm/km^2 in 2000. In contrast, the range in 2018 was from 0 to 544.406 mm/k^2, with an average of 229 mm/km^2, marking an approximate 60% increase. Geographically, the water yield distribution was fragmented in 2000. However, the 2018 data displayed a distinct hierarchical boundary, extending outward from the central three rivers source region, primarily Nangqian County in the Yushu Tibetan Autonomous Prefecture, as the core. Conversely, the western regions, including western Zhiduo County and western Golmud City, and the eastern areas, comprising Maqin County, Gande County, Darri County, Jiuzhi County, and Banma County in the Guoluo Tibetan Autonomous Prefecture, recorded notably lower yields (Fig. 7.2).

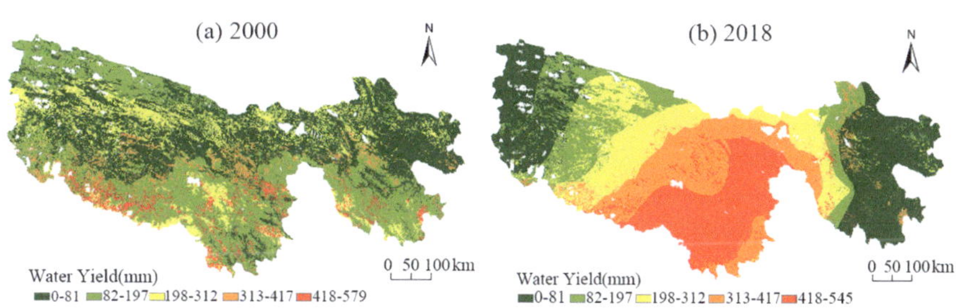

Fig. 7.2 Spatial distribution of water yield

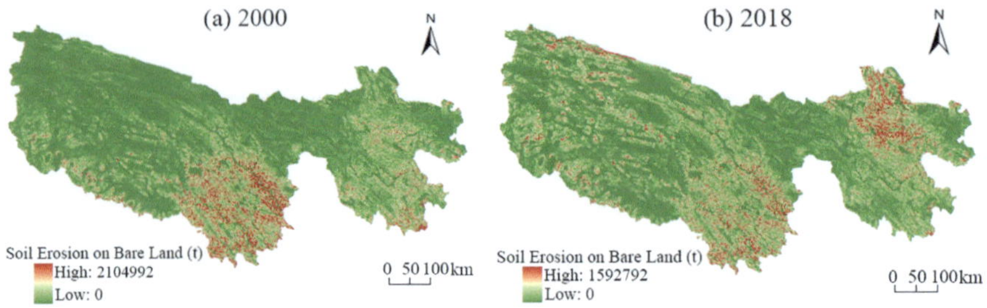

Fig. 7.3 Spatial distribution of bare soil erosion

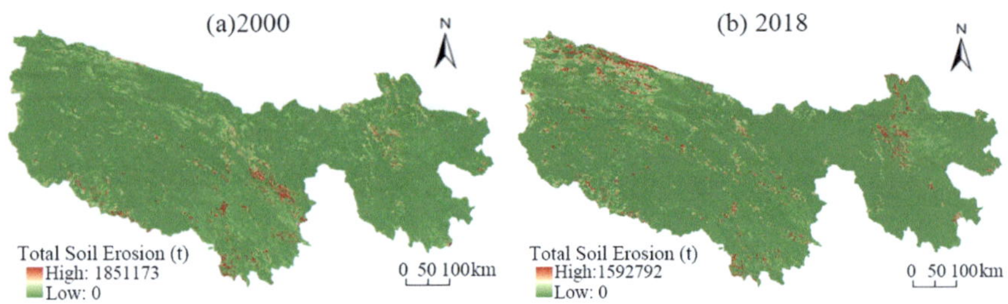

Fig. 7.4 Spatial distribution of total soil erosion

7.3.1.3 Soil and Water Conservation Simulation

Figure 7.3 indicates that bare soil erosion peaked at 2.10499 million tons per square kilometer in 2000, with a total of 14,344 million tons nationwide. By 2018, it had increased to 1.59279 million tons per square kilometer, reaching 16,507 million tons. Although the maximum erosion per grid cell decreased, the total bare soil erosion increased regionally. From 2000 to 2018, severe soil erosion was concentrated in densely populated southeastern areas, particularly in the Tongtian River Basin (Fig. 7.4). In 2000, the three rivers source region recorded the highest soil erosion at 1.85117 million tons per square kilometer, totaling 215 million tons nationwide. By 2018, the peak erosion per square kilometer reached 1.59279 million tons, with a total of 295 million tons. Overall, there is a positive correlation between total soil erosion and bare soil erosion, with higher values mainly occurring along major riverbanks. Additionally, low vegetation coverage in northwestern regions also contributes to severe soil erosion.

7.3.1.4 Habitat Quality Simulation

Overall, the highest relative habitat degradation level in 2000 was 10.397%, while the three rivers source region recorded a slightly higher rate of 11.150% in 2018. However, the overall trend shows a gradual decline in relative habitat degradation

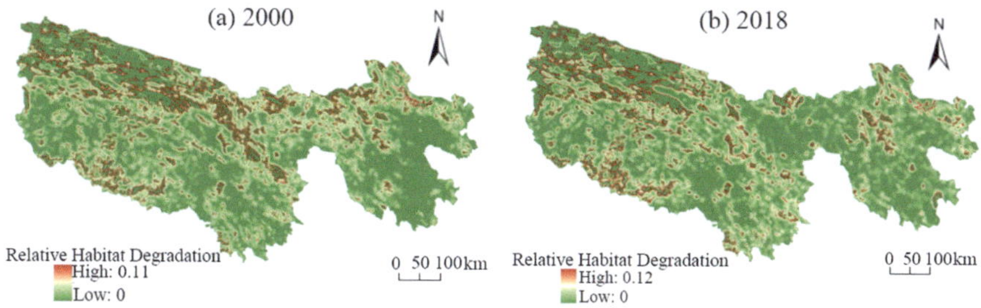

Fig. 7.5 Spatial distribution of relative habitat degradation level

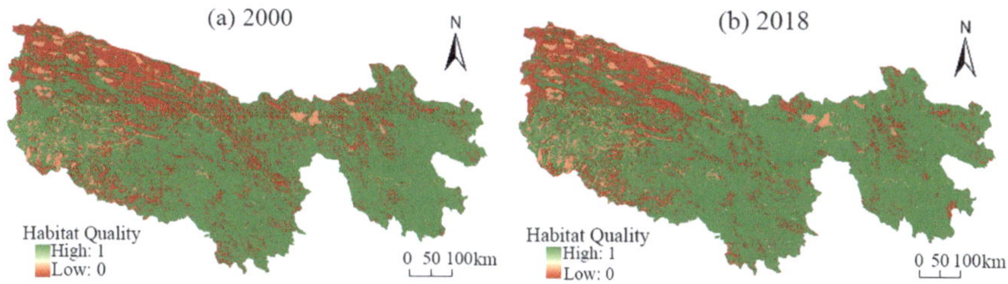

Fig. 7.6 Spatial distribution of habitat quality

(Fig. 7.5), with the regional average decreasing from 0.008 in 2000 to 0.007 in 2018, and habitat quality improving from 0.891 to 0.907. Spatially, areas with higher relative habitat degradation are scattered throughout the study area (Fig. 7.6). Regarding threat factors, analysis of land use patterns in the three rivers source region reveals that unused land and road infrastructure pose the greatest threats to habitat quality. The highest relative degradation levels occur at road–land interface areas, while uninhabited regions in the northwest exhibit weaker habitat quality. This paradox may stem from the harsh Gobi desert conditions in these areas, which hinder species reproduction and consequently result in lower biodiversity.

7.3.2 Graded Evaluation of Ecological Barrier Function Indicators

Based on the quantitative synthesis of principal components, the ecological barrier function efficiency index for the three rivers source region was classified using a uniform grading system, with scores ranging from 1 to 5. The rankings, from lowest to highest, are: severe, poor, moderate, good, and excellent.

The carbon sequestration and oxygen release functions, species conservation functions, and water source conservation functions were also graded using this same system.

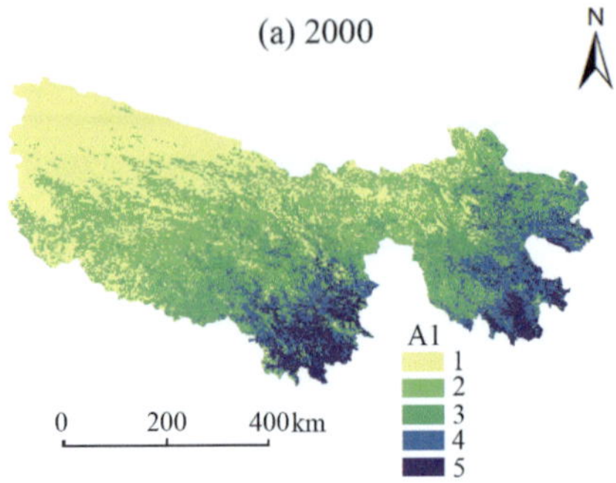

Fig. 7.7 Hierarchical classification of principal component A1. *Note* AI is carbon sequestration and oxygen release; Levels 1 to 5 in the legend indicate poor, fair, average, good, and excellent, respectively

Soil erosion was categorized into five levels, with scores ranging from highest to lowest as: severe, intense, moderate, mild, and slight. A severe score indicates significant soil erosion and low soil conservation efficiency.

The ecological barrier function efficiency index for the three rivers source region and the sub-function values derived from the four principal components were spatially mapped to each geographical sample, resulting in the creation of zoning maps for principal components A1, A2, A3, A4, and ecological barrier function levels (Fig. 7.7).

As shown in Fig. 7.8, the efficiency of carbon sequestration and oxygen release presents a spatial distribution, with higher values in the southeast and lower values in the northwest. The areas with poor efficiency are concentrated in the northwest, while those with excellent efficiency are concentrated in the southeast. From 2000 to 2018, the carbon sequestration and oxygen release efficiency in the three rivers source region showed an upward trend. All areas had lower values in the northwest and higher values in the southeast, and the eastern region had the most significant increase.

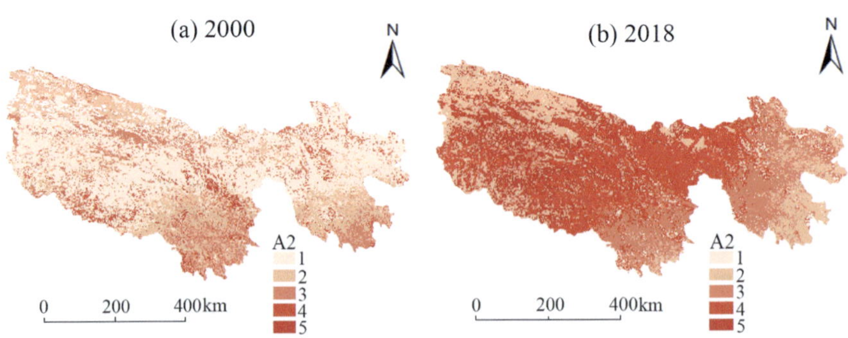

Fig. 7.8 Principal component A2 (species conservation) classification map. *Note* Levels 1 to 5 in the legend indicate poor, fair, average, good, and excellent, respectively

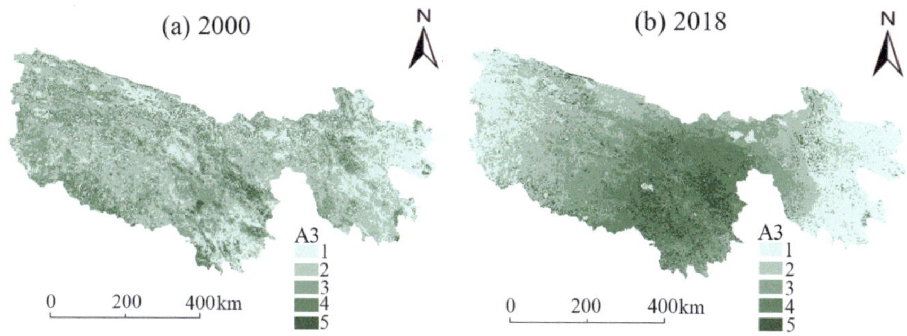

Fig. 7.9 Hierarchical classification of principal component A3 (water source conservation). *Note* Levels 1 to 5 in the legend indicate poor, fair, average, good, and excellent, respectively

As shown in Fig. 7.9, there are obvious spatial and temporal differences in species conservation functions. Compared with 2000, the number of regions with excellent species conservation benefits in 2018 has increased significantly, indicating that the species conservation work in the three rivers source region has achieved great results.

The spatial distribution of water conservation benefits largely corresponds with the spatial distribution pattern of water yield. In 2000, the distribution was scattered, and the overall water conservation benefits were lower than in 2018. By 2018, the grade differentiation was pronounced, with water conservation benefits diminishing progressively from the center of Yushu and Chengdu towards the surrounding areas (Fig. 7.10).

The effectiveness of soil conservation shows a strong correlation with the watershed scale, with the most prominent correlation being observed in 2000. The Tongtian River Basin experienced severe soil erosion, leading to the least effective soil conservation results.

By 2018, areas with severe soil erosion had significantly decreased. However, Xinghai County and Tongde County still faced serious soil erosion problems.

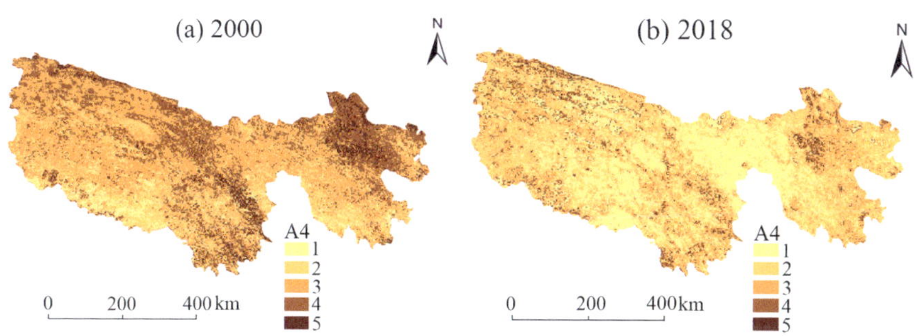

Fig. 7.10 Principal component A4 (soil erosion) classification map. *Note* The legend shows levels 1 to 5 is extremely slight, slight, moderate, severe, and very severe

Notably, the soil erosion along the Tongtian River showed significant improvement, resulting in enhanced soil conservation effectiveness.

7.3.3 *Comprehensive Ecological Barrier Function and Benefit Evaluation*

7.3.3.1 Construction of Comprehensive Ecological Barrier Function and Benefit Evaluation Index System

Through principal component analysis, the comprehensive model of principal components was constructed using the proportion of each component's eigenvalue λ to the total extracted eigenvalues as weights. This quantitative analysis of the ecological barrier function in the three rivers source region yielded the ecological barrier function benefit index I, as shown in Table 7.12.

$$I = \frac{\lambda_1}{\lambda_1 + \lambda_2 + \lambda_3 + \lambda_4} A1 + \frac{\lambda_2}{\lambda_1 + \lambda_2 + \lambda_3 + \lambda_4} A2 + \frac{\lambda_3}{\lambda_1 + \lambda_2 + \lambda_3 + \lambda_4} A3 + \frac{\lambda_4}{\lambda_1 + \lambda_2 + \lambda_3 + \lambda_4} A4 \quad (7.8)$$

In the formula, I denotes the ecological barrier function benefit index, while λ_1 to λ_4 represent the eigenvalues of the principal components A1 to A4.

Table 7.12 Ecological barrier function benefit index of three rivers source region (partial)

Geographic sample	I (2000)	I (2018)
1	4.648	1.380
2	1.366	0.150
3	0.009	0.650
4	1.691	1.138
5	1.666	1.729
6	– 0.140	– 0.494
7	– 1.318	– 2.645
8	1.391	1.323
9	– 1.990	– 1.687
10	1.155	1.175

7.3.3.2 Results of Comprehensive Ecological Barrier Function and Benefit Evaluation

As shown in Fig. 7.11, the central-southern and eastern regions of the three rivers source region exhibit excellent ecological barrier functions. The arid and semi-arid zones of China are demarcated here from west to east, with arid areas mainly distributed in regions with poor ecological barrier performance. The boundary between the inland and outflow zones runs from northeast to southwest, and the inland zone consists of areas with suboptimal ecological barrier function. The demarcation between poor and weak ecological barrier performance roughly coincides with the dry–wet boundary and the inland–outflow zone boundary. The Yangtze River source area shows weaker ecological barrier functions compared to the Yellow River source area and the Lancang River source area, while the Hoh Xil region demonstrates poorer ecological barrier performance than other areas. Overall, ecological barrier functions gradually strengthen from northwest to southeast.

The analysis of various ecological benefits reveals distinct regional patterns. In 2000, the central and southern regions displayed strong carbon sequestration capacity but faced low biodiversity and severe soil erosion, whereas the eastern regions showed robust soil conservation. By 2018, all ecological benefits had improved. The ecological barrier function in China's semi-arid and semi-humid zones generally corresponded to the moderate and good functional zones, respectively, while the non-monsoon and monsoon regions corresponded to the good and excellent functional zones.

A comparative analysis of ecological barrier functions and their sub-functional benefit classifications between 2000 and 2018 reveals significant spatiotemporal variations. While the soil conservation capacity declined, all other functional benefits showed marked improvement, which is consistent with previous regional studies.

Considering the historical evolution of the Three Rivers Source Region, its ecological barrier system has inherent fragility. However, extensive development and construction activities since the reform and opening-up period have severely

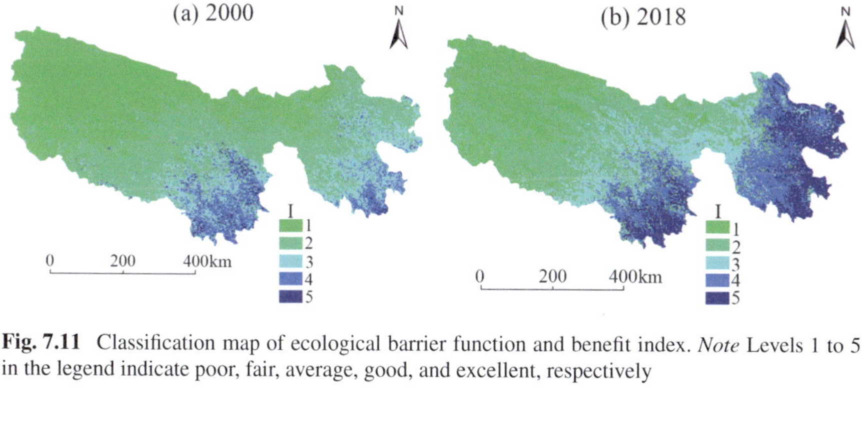

Fig. 7.11 Classification map of ecological barrier function and benefit index. *Note* Levels 1 to 5 in the legend indicate poor, fair, average, good, and excellent, respectively

compromised the ecological integrity of the three rivers source region, exacerbating the decline in ecological benefits.

Since 2000, a series of environmental protection measures, including the Grain for Green Program and the Three Rivers Source Region Nature Reserve Project, have been implemented. Consequently, the comprehensive evaluation results in 2018 showed significant improvements in ecological barrier functions and benefits, demonstrating the remarkable effectiveness of nature conservation efforts in the three rivers source region.

7.3.4 Analysis of Driving Forces of Ecological Barrier Function

7.3.4.1 Detection of Core Driving Forces of Ecological Barrier Function

Through factor analysis, we quantitatively assess the impact of various driving factors on the functions of ecological barriers, as illustrated in Table 7.13. Risk analysis also helps to pinpoint the directional effects of these driving factors. The ranking of contribution rates for the Ecological Barrier Function Benefit Index (I) closely corresponds with the A1 category of the primary urban area (carbon sequestration and oxygen release), which shows the highest contribution rate to I. The data presented in the table suggest that indicators X5 (annual precipitation), X6 (annual temperature), and X7 (relative humidity) significantly contribute to all ecological barrier functions. This implies that natural climatic factors, including precipitation, temperature, and humidity, are key drivers for both the overall ecological barrier benefits and their specific functional aspects.

7.3.5 Core Driving Force Analysis

The analysis of spatial heterogeneity in ecological barrier function indices indicated that all p-values were statistically significant. The contribution rates were ranked as follows: Annual Precipitation (0.180) > Annual Temperature (0.069) > Relative Humidity (0.068).

For carbon sequestration and oxygen release, the order was Precipitation (0.188) > Annual Temperature (0.073) > Relative Humidity (0.071).

Species conservation showed Annual Temperature (0.010) > Relative Humidity (0.009) > Clay Content (0.007).

Water conservation ranked Precipitation (0.032) > Temperature (0.024) > Landform Type (0.021).

Soil conservation demonstrated Precipitation (0.011) > Population Density (0.070) > Temperature (0.007) (Table 7.14).

Table 7.13 Detection results (q-values) of ecological barrier functions and subfunctions

Metric	I	Sort	A1	Sort	A2	Sort	A3	Sort	A4	Sort
X5	0.180	1	0.188	1	0.005	4	0.032	1	0.011	1
X6	0.069	2	0.073	2	0.010	1	0.024	2	0.007	3
X7	0.068	3	0.071	3	0.009	2	0.006	12	0.000	14
X9	0.058	4	0.060	4	0.004	5	0.017	5	0.002	10
X10	0.049	5	0.051	5	0.007	3	0.009	9	0.000	13
X4	0.043	6	0.044	7	0.000	12	0.021	3	0.006	4
X15	0.043	7	0.044	6	0.002	6	0.014	6	0.007	2
X3	0.026	8	0.027	8	0.000	14	0.005	13	0.002	8
X14	0.020	9	0.020	9	0.001	10	0.007	10	0.003	7
X13	0.019	10	0.020	10	0.000	11	0.011	8	0.006	5
X1	0.016	11	0.017	11	0.001	8	0.017	4	0.005	6
X12	0.012	12	0.013	12	0.000	15	0.001	15	0.002	9
X11	0.011	13	0.011	13	0.002	7	0.007	11	0.001	12
X8	0.009	14	0.010	14	0.001	9	0.012	7	0.001	11
X2	0.002	15	0.002	15	0.000	13	0.001	14	0.000	15

Table 7.14 Core driving force detection results (q-values)

	Core drivers		
	1	2	3
Ecological barrier function (I)	X5 Average annual precipitation 0.180	X6 Annual average temperature 0.069	X7 Relative humidity 0.068
Carbon sequestration and oxygen release (A1)	X5 Average annual precipitation 0.188	X6 Annual average temperature 0.073	X7 Relative humidity 0.071
Species Conservation (A2)	X6 Annual average temperature 0.010	X7 Relative humidity 0.009	X10 Clay content 0.007
Water conservation (A3)	X5 Average annual precipitation 0.032	X6 Annual average temperature 0.024	X4 geomorphic type 0.021
Soil erosion (A4)	X5 Average annual precipitation 0.011	X15 Density of population 0.070	X6 Annual average temperature 0.007

Note The top three q-values are the core drivers

7.3.5.1 Driving Force Interaction Analysis

The interaction detector module is utilized to ascertain the relative significance of driving factors influencing the benefits of ecological barrier functions, and to assess the impact of these factors on the research subject through the interaction between them (Table 7.15).

The interactions among factors are predominantly nonlinear, while the rest exhibit bivariate effects. There is no independent interaction between the bivariate factors. Analysis of interaction patterns reveals that, among the driving forces affecting ecological barrier functions, the interaction between annual average precipitation (X5) and relative humidity has the strongest explanatory power. Therefore, precipitation and relative humidity have the most significant influence on the ecological barrier function index.

Among the interaction effects of carbon sequestration and oxygen release benefits, the interaction between annual precipitation (X5) and relative humidity demonstrated the highest explanatory power. The interactions between altitude (X1), slope direction (X2), and other factors were predominantly non-linear enhancements, suggesting that altitude and slope direction significantly amplify the effects of other factors.

Table 7.15 Main results of interaction detection

Nonlinear enhancement				Dual factor enhancement		
Xs to I		Xs vs A1		XS vs A2	Xs vs A3	Xs to A4
X10 ∩ X8	X1 ∩ X11	X10 ∩ X8	X1 ∩ X11	X10 ∩ X9	X8 ∩ X10	X15 ∩ X13
X10 ∩ X2	X1 ∩ X12	X10 ∩ X2	X1 ∩ X12	X10 ∩ X8	X13 ∩ X10	X15 ∩ X5
X9 ∩ X1	X1 ∩ X13	X9 ∩ X1	X1 ∩ X13		X8 ∩ X9	X12 ∩ X13
X9 ∩ X8	X2 ∩ X3	X9 ∩ X8	X2 ∩ X3		X9 ∩ X3	X12 ∩ X5
X9 ∩ X2	X2 ∩ X4	X9 ∩ X2	X2 ∩ X4		X9 ∩ X15	X13 ∩ X5
X8 ∩ X1	X2 ∩ X6	X8 ∩ X1	X2 ∩ X6		X3 ∩ X4	X11 ∩ X12
X8 ∩ X2	X2 ∩ X7	X8 ∩ X2	X2 ∩ X7		X3 ∩ X5	X9 ∩ X8
X8 ∩ X3	X2 ∩ X11	X8 ∩ X3	X2 ∩ X11		X3 ∩ X12	X3 ∩ X12
X8 ∩ X11	X2 ∩ X12	X8 ∩ X11	X2 ∩ X12		X3 ∩ X13	X3 ∩ X15
X8 ∩ X14	X2 ∩ X13	X8 ∩ X14	X2 ∩ X13		X3 ∩ X15	X4 ∩ X12
X8 ∩ X15	X2 ∩ X14	X8 ∩ X15	X2 ∩ X14		X4 ∩ X5	X10 ∩ X12
X1 ∩ X14	X2 ∩ X15	X1 ∩ X14	X2 ∩ X15		X14 ∩ X12	
X1 ∩ X15	X3 ∩ X7	X1 ∩ X15	X3 ∩ X7		X14 ∩ X5	
X1 ∩ X2	X3 ∩ X11	X1 ∩ X2	X3 ∩ X11		X15 ∩ X13	
X1 ∩ X3	X4 ∩ X6	X1 ∩ X3	X4 ∩ X6		X15 ∩ X5	
X1 ∩ X4	X4 ∩ X11	X1 ∩ X4	X4 ∩ X11			
X1 ∩ X5	X4 ∩ X13	X1 ∩ X5	X4 ∩ X13			
X1 ∩ X6	X11 ∩ X13	X1 ∩ X6	X11 ∩ X13			
X1 ∩ X7	X6 ∩ X7	X1 ∩ X7	X6 ∩ X7			

In the interaction effects of conservation benefits on species, only clay content, silt content, and sand content exhibited dual-factor enhancement, while the others showed nonlinear enhancement. This indicates that soil texture exerts a stronger interactive influence on conservation benefits than individual factors, whereas the remaining factors demonstrate more pronounced mutual reinforcement effects.

In the interaction of factors influencing soil and water conservation benefits, over 10 driving forces exhibit nonlinear enhanced interactions with one another. Among these, slope direction, annual average relative humidity, sand content, and population point density also demonstrate nonlinear enhanced interactions with each driving force.

7.3.6 Carbon Sink Function Analysis of Grassland in the Three Rivers Source Region

The carbon sequestration and oxygen production capabilities of the three rivers source region were evaluated through model simulations to assess its productivity, coverage, and carbon sink functions. Figure 7.12 depicts the spatial distribution of the annual mean net primary productivity (NPP) and the inter-annual variation rates of grasslands in the three rivers source region.

The annual mean NPP of the grasslands in the region is 401.2 gC/m^2, showing a gradual increase from west to east. The highest values are observed in eastern counties such as Banma, Jiuzhi, Gande, and Henan. In contrast, lower values are noted in western areas, including Golmud City, Zhiduo County, and Qumalai County.

The total annual mean NPP of the grasslands amounts to 95.8 TgC. The inter-annual variation rate of grassland NPP is—1.7 gC/m^2, resulting in a total reduction of 7.1 TgC.

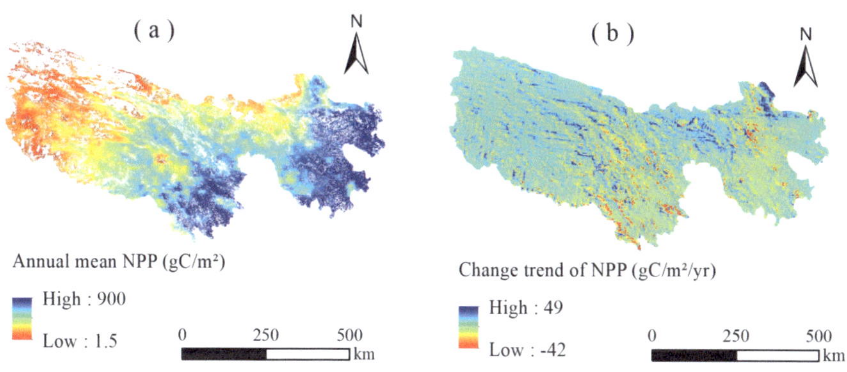

Fig. 7.12 Spatial distribution of grassland annual mean NPP (**a**), and change trend in the three rivers source region

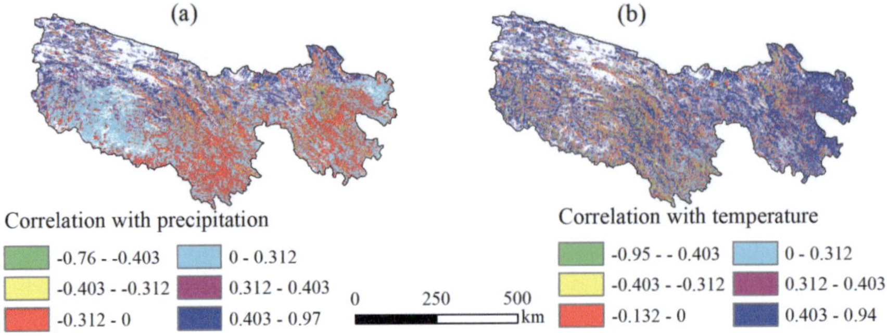

Fig. 7.13 Spatial distribution of the correlation coefficient between NPP and annual precipitation/average temperature in three rivers source region grassland

Spatially, Jiuzhi County, Darri County, and Maduo County show significant increases in NPP. These increases are attributed to their more variable hydrothermal conditions and relatively successful ecological engineering projects, leading to dramatic changes in grassland NPP.

Figure 7.13 depicts the spatial distribution of the partial correlation coefficient between grassland net primary productivity (NPP) and annual precipitation/average temperature in the three rivers source region. The spatial pattern shows stronger correlations in the northwest than in the southeast, probably because the northwestern region receives more precipitation, which significantly affects plant growth. The average correlation coefficient between NPP and temperature across the study area is 0.65, indicating a positive correlation. Although most of the western and central regions show positive correlations with temperature, the NPP in Nangqian County, Yushu County, Banma County, and Jiuzhi County mainly shows negative correlations with the annual average temperature.

The average soil organic carbon density of grassland in the three rivers source region is 21.1 kg C/m^2. Given the presence of extensive alpine meadows and alpine pastures within the source area, these grassland types demonstrate a better organic carbon storage capacity than other vegetation types. The total organic carbon reserves in the grassland soil of the three rivers source region reach 5029.1 Tg C.

The annual net ecosystem productivity (NEP) of grasslands in the three rivers source region averages 104.1 g C /m^2, indicating overall carbon sequestration, with a total annual carbon sink of 24.8 Tg C.

Most western areas of the three rivers source region exhibit positive NEP values. Tongde County shows the highest annual carbon sink at 187.0 g of carbon per square meter (g C/m^2), while Yushu County has the lowest at 69.3 g C/m^2.

However, regions including Chengduo County, Nangqian County, Yushu County, Dari County, Jiuzhi County, and Banma County exhibit relatively weak carbon sequestration. Notably, these areas make up a larger proportion of the region's total carbon-sink area, with annual carbon emissions ranging from 97 to 296 g C/m^2 compared to other counties.

7.4 Conclusions

Upon comparing and analyzing the data from 2000 and 2018, this study comprehensively uncovers the spatio-temporal dynamic change laws and internal driving forces behind the ecological barrier function in the three rivers source region. The primary conclusions are as follows:

(1) The total carbon storage in the three rivers source region increased from 753.7 Tg C in 2000 to 781.2 Tg C in 2018, indicating a net increase of 27.5 Tg C. The carbon density reached a peak of 3,396 Mg/km^2. Structurally, forest land, grassland, and unused land consistently dominated carbon storage, collectively accounting for over 96% of the total.

Spatially, high-carbon storage areas remained concentrated in the southeastern forest and high-coverage grassland regions (including parts of Luozhou and southern Yushu Prefecture), while low-carbon storage areas (<120 Mg/km^2) were clustered in the northwestern Kekexili region. The overall "southeast high, northwest low" gradient pattern has remained fundamentally unchanged.

(2) Water yield capacity emerged as the most significant improvement in ecological services during the study period. The regional average water yield soared from 137 to 229 mm, a remarkable increase of 67.2%. Meanwhile, the spatial distribution pattern changed from chaotic fragmentation in 2000 to a highly structured configuration by 2018. This new pattern was centered around the central region of Yushu Tibetan Autonomous Prefecture (Nangqian County, Yushu City, and Chengduo County), with values gradually decreasing outward. These findings indicate that the water conservation capacity has not only achieved substantial quantitative growth but also shown remarkable optimization in spatial coordination.

(3) Soil erosion remains the sole indicator showing a deteriorating trend. The total area affected by bare-soil erosion has expanded from 1,434,479.405 million tons to 1,650,740.271 million tons, and the total volume of soil erosion has increased from 215,771.375 million tons to 295,158.724 million tons, which reflects the severe overall situation of soil erosion.

Spatially, erosion hotspots have concentrated in the densely populated southeastern regions, especially in the Tongtian River Basin. However, data also indicates a reduction in the intensity of soil erosion in some undeveloped areas of the northwest and locally managed riverbank zones, demonstrating the localized effectiveness of remediation measures.

(4) Habitat quality has been effectively restored, with the regional average habitat quality index increasing from 0.891 to 0.907. Meanwhile, the average relative habitat degradation level dropped from 0.008 to 0.007. The stable ranking of habitat quality endures: forest land > grassland > water area > unused land > cultivated land > construction land. However, due to urbanization and the expansion of transportation networks, the relative habitat degradation in urban areas, along major roads, and around cultivated regions has remained consistently high

(up to 11.15%). These ecologically vulnerable areas demand focused attention for future biodiversity conservation endeavors.

A comprehensive analysis has revealed that the three rivers source region underwent significant spatiotemporal restructuring of its ecological barrier functions from 2000 to 2018. During this period, the functions of carbon sequestration, oxygen release, water conservation, and species preservation were systematically strengthened. However, soil and water conservation remained the critical vulnerability restricting regional ecological security.

Analysis of the driving factors identified annual precipitation ($q = 0.180$), average temperature ($q = 0.069$), and relative humidity ($q = 0.068$) as the primary natural drivers affecting the spatial differentiation of ecological functions.

The integrated evaluation index system and dynamic assessment results developed in this study provide robust scientific evidence and data support for precise ecological management and the effectiveness evaluation of the Three Rivers Source Region National Park.

References

Ali, D. A., Deininger, K., & Monchuk, D. (2020). Using satellite imagery to assess impacts of soil and water conservation measures: Evidence from Ethiopia's Tana-Beles watershed. *Ecological Economics, 169*, 106512.

Cheng, Y. (2019). *Study on the change of water conservation capacity and grassland ecological compensation in SanjiangyuanNational park based on InVEST model*. Yunnan University of Finance and Economics.

Dixon, R. K., Solomon, A. M., Brown, S., Houghton, R. A., Trexier, M. C., & Wisniewski, J. (1994). Carbon pools and flux of global forest ecosystems. *Science, 263*, 185–190.

Goldstein, J. H., Caldarone, G., Duarte, T. K., Ennaanay, D., Hannahs, N., Mendoza, G., Polasky, S., Wolny, S., & Daily, G. C. (2012). Integrating ecosystem-service tradeoffs into land-use decisions. *Proceedings of the National Academy of Sciences, 109*, 7565–7570.

Herrmann, S., Dabbert, S., & Schwarz-von Raumer, H. G. (2003). Threshold values for nature protection areas as indicators for bio-diversity—a regional evaluation of economic and ecological consequences. *Agriculture, Ecosystems & Environment, 98*, 493–506.

Hudak, A. T., Strand, E. K., Vierling, L. A., Byrne, J. C., Eitel, J. U. H., Martinuzzi, S., & Falkowski, M. J. (2012). Quantifying aboveground forest carbon pools and fluxes from repeat LiDAR surveys. *Remote Sensing of Environment, 123*, 25–40.

Ivits, E., Horion, S., Fensholt, R., & Cherlet, M. (2014). Drought footprint on European ecosystems between 1999 and 2010 assessed by remotely sensed vegetation phenology and productivity. *Global Change Biology, 20*, 581–593.

Liu, W. (2012). *On the theoretical foundation and Developmental Bath of China ecological civilization*. Hebei Normal University.

Loomis, J., Kent, P., Strange, L., Fausch, K., & Covich, A. (2000). Measuring the total economic value of restoring ecosystem services in an impaired river basin: Results from a contingent valuation survey. *Ecological Economics, 33*, 103–117.

Pan, T., Wu, S., Dai, E., & Liu, Y. (2013). Spatiotemporal variation of water source supply service in three rivers source area of chi-na based on InVEST model. *Chinese Journal of Applied Ecology, 24*, 183–189.

Petit, C. C., & Lambin, E. F. (2001). Integration of multi-source remote sensing data for land cover change detection. *International Journal of Geographical Information Science, 15*, 785–803.
Redhead, J. W., Stratford, C., Sharps, K., Jones, L., Ziv, G., Clarke, D., Oliver, T. H., & Bullock, J. M. (2016). Empirical validation of the InVEST water yield ecosystem service model at a national scale. *Science of the Total Environment, 569–570*, 1418–1426.
Wang, W., Sun, T., Wang, J., Fu, Q., & An, C. (2019). Annual dynamic monitoring of regional ecosystem service ValueBased on multi-source remote sensing data: A case of central plains urban agglomeration region. *Scientia Geographica Sinica, 39*, 680–687.
Wei, W., Zhang, K., & Zhou, J. (2020). Review and prospect of human-land relationship in three river headwaters region: Based on the perspective of people, events, time and space. *Advances in Earth Science, 35*, 26–37.
Zhang, W. (2006). *Organic carbon accumulation and carbon cycling simulating in Typical wetland ecosystems*. Huazhong Agricultural University.

MIX
Papier aus verantwortungsvollen Quellen
Paper from responsible sources
FSC® C105338

If you have any concerns about our products,
you can contact us on
ProductSafety@springernature.com

In case Publisher is established outside the EU,
the EU authorized representative is:
Springer Nature Customer Service Center GmbH
Europaplatz 3, 69115 Heidelberg, Germany

Printed by Libri Plureos GmbH
in Hamburg, Germany